Dictionary
of
Food Ingredients

Robert S. Igoe

Senior Technical Representative
Kelco Division of Merck & Co., Inc.

VNR VAN NOSTRAND REINHOLD
——————— NEW YORK

Library of Congress Catalog Card Number 88-38855
ISBN 0-442-31927-4

Printed in the United States of America

Van Nostrand Reinhold
115 Fifth Avenue
New York, New York 10003

Chapman & Hall
2-6 Boundary Row
London SE1 8HN, England

Thomas Nelson Australia
102 Dodds Street
South Melbourne, Victoria 3205, Australia

Nelson Canada
1120 Birchmount Road
Scarborough, Ontario M1K 5G4, Canada

16 15 14 13 12 11 10 9 8 7 6 5 4

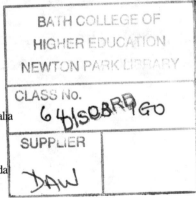

Library of Congress Cataloging-in-Publication Data

Igoe, Robert S.
  Dictionary of food ingredients/Robert S. Igoe.—2nd ed.
   p. cm.
  Bibliography: p.
  ISBN 0-442-24002-3
  1. Food—Composition—Dictionaries. I. Title.
TX551.I26   1989
641.1′03′21—dc19   88-38855
                                                     CIP

Contents

Preface

The *Dictionary of Food Ingredients* is a reference source useful for anyone involved with foods. It is a source of information on approved food ingredients, ingredient categories, and ingredients listed under the Code of Federal Regulations.

Within the Ingredients section, the alphabetized ingredients are concisely defined according to their functions, properties, and applications. Chemical formulations and usage levels are provided where applicable. This second edition includes additional ingredients and provides expanded information for many of the previously listed ingredients. The Ingredients Categories section groups principal food ingredients, providing collective information on the properties and relation of the respective component ingredients. Charts that identify key properties and features of the ingredients are a new addition. A new section titled Substances for Use in Foods Listed Under Title 21 of the Code of Federal Regulations includes six parts listing food ingredients and the respective United States approved status. The Bibliography section indicates the sources from which the reference material was obtained.

I would like to express my appreciation to the companies and authors that provided information on food ingredients defined in this book.

R. S. Igoe

Ingredients

Acacia   See **Arabic**.

Acesulfame-K   A non-nutritive sweetener, also termed acesulfame potassium. It is a white, crystalline product that is 200 times sweeter than sucrose. It is not metabolized in the body. It is relatively stable as a powder and in liquids and solids which may be heated. Acesulfame-K is approved for use in dry food products.

Acesulfame Potassium   See **Acesulfame-K.**

Acetic Acid $C_2H_4O_2$. An acid produced chemically from the conversion of alcohol to acetaldehyde to acetic acid. It is the principal component of vinegar which contains not less than 4 g of acetic acid in 100 cm³ at 20°C. The approved salts include sodium acetate, calcium acetate, sodium diacetate, and calcium diacetate. It is used as a preservative, acidulant, and flavoring agent in catsup, mayonnaise, and pickles. It can be used in conjunction with leavening agents to release carbon dioxide from sodium bicarbonate.

Acetic Acid Ester of Monoglyceride   See **Acetylated Monoglyceride**.

Acetic Acid, Glacial   See **Glacial Acetic Acid**.

Acetic Anhydride $C_4H_6O_3$. An esterifier for food starch; also used in combination with adipic anhydride.

Acetone Peroxide A dough conditioner, maturing and bleaching agent that is a mixture of monomeric and linear dimeric acetone peroxides which are strong oxidizing agents. It is used to age and bleach flour.

Acetylated Monoglyceride An emulsifier manufactured by the inter-esterification of edible fats with triacetin and in the presence of catalysts or by the direct acetylation of edible monoglycerides with acetic anhydride without the use of catalysts. It is characterized by sharp melting points, stability to oxidative rancidity, film forming, stabilizing, and lubricating properties. It is used as protective coatings for meat products, nuts, and fruits to improve their appearance, texture, and shelf life. The coatings are applied by spraying, panning, and dipping. It is used in cake shortening and fats for whipped topping to enhance the aeration and improve foam stabilization. It is found in dry-mix whipped topping.

Acetylated Tartaric Acid Monoglyceride See **Diacetyl Tartaric Acid Esters of Mono- and Diglycerides**.

Acetyl Tartrate Mono- and Diglyceride See **Diacetyl Tartaric Acid Esters of Mono- and Diglycerides**.

Acid Calcium Phosphate See **Monocalcium Phosphate**.

Acid Casein The principal milk protein which is prepared from skim milk by precipitation with an acid, such as lactic, sulfuric, or hydrochloric acid, to lower the pH of the milk to 4.4 to 4.7. Caseins are identified according to the type of acid used, but in this form have little utility in foods, though they are used to some extent in cereal and bread fortification. Neutralization of the caseins yields the salts of which sodium and calcium caseinates are the most common. See **Casein**.

Acid-Modified Corn Starch See **Corn Starch, Acid-Modified**.

Acidophilus A bacterial starter culture used to develop flavor in dry sausage, pork roll, bologna, and salami at a level of 0.5 percent. It is used in the production of acidophilus milk.

Acid Sodium Pyrophosphate See **Sodium Acid Pyrophosphate**.

Acrolein This is used in the ether etherification of food starch up to

0.6 percent and for the esterification and etherification of food starch up to 0.3 percent with vinyl acetate up to 7.5 percent.

Adipic Acid $C_6H_{10}O_4$. An acidulant and flavoring agent. It is characterized as stable, nonhygroscopic, and slightly soluble, with a water solubility of 1.9 g per 100 ml at 20°C. It has a pH of 2.86 at 0.6 percent usage level at 25°C. It is used in powdered drinks, beverages, gelatin desserts, lozenges, and canned vegetables. It is also used as a leavening acidulant in baking powder. It can be used as a buffering agent to maintain acidities within a range of pH 2.5 to 3.0. It is occasionally used in edible oils to prevent rancidity.

Adipic Anhydride   An esterifier for food starch in combination with acetic anhydride.

Agar   A gum obtained from red seaweeds of the genera *Gelidium*, *Gracilaria*, and *Eucheuma*, class Rhodophyceae. It is a mixture of the polysaccharides agarose and agaropectin. It is insoluble in cold water, slowly soluble in hot water, and soluble in boiling water forming a gel upon cooling. The gels are characterized as being tough and brittle, setting at 32° to 40°C and melting at 95°C. A rigid, tough gel can be formed at 0.5 percent. Agar mainly functions in gel formation because of its range between melting and setting temperatures, being used in piping gels, glazes, icings, dental impression material, and microbiological plating. Typical use levels are 0.1 to 2.0 percent.

Agar-Agar   See **Agar**.

Albumin   Any of several water-soluble proteins that are coagulated by heat and are found in egg white, blood serum, and milk. Milk albumin is termed lactalbumin and milk albuminate and it contains 28 to 35 percent protein and 38 to 52 percent lactose. It is used as a binder in imitation sausage, soups, and stews.

Aldehyde C-16   See **Ethyl Methyl Phenyl Glycidate**.

Aldehyde C-18   See **(Gamma)-Nonalactone**.

Algin   A gum, a term used for derivatives of alginic acid which are obtained from brown seaweed genera, such as *Macrocystis pyrifera*. The derivatives are sodium, ammonium, and potassium alginates of which the sodium salt is most common. They are used to provide thickening, gelling, and binding. A derivative designed for improved acid and calcium stability is propylene glycol alginate. The algins are soluble in cold water and form non-thermoreversible gels in reaction with calcium ions and under acidic conditions. Algin is used in ice cream, icings, puddings, dessert gels, and fabricated fruit.

Alginate   A gum derived from alginic acid that is used to provide thickening, gelling, and binding. See **Algin**.

Alginic Acid $(C_6H_8O_6)_n$. The acidic, insoluble form of algin that is a white to yellowish fibrous powder obtained from the brown seaweed genera such as *Macrocystis pyrifera*. The derivatives are soluble and include sodium, potassium, and ammonium alginate and propylene glycol alginate. It is used as a tablet disintegrant and as an antacid ingredient.

All-Purpose Flour   A flour that is intermediate between long-patent flours (bread flour) containing more than 10.5 percent protein and 0.40 to 0.50 percent ash and short-patent flours (cake flour) which generally contain less than 10 percent protein and less than 0.40 percent ash. It is made from hard or soft wheat and is used in baking and in gravies. It is also termed family flour.

Allspice   A spice made from the dried, nearly ripe berries of *Pimenta officinalis*, a tropical evergreen tree. It has an aroma and flavor resembling that of a blend of nutmeg, cinnamon, and cloves. For labeling purposes, allspice refers to the spice of Jamaican origin. It is used in fruit pies, cakes, mincemeat, plum pudding, soups, and sauces.

Allura® Red AC   See **FD&C Red #40**.

Allyl Anthranilate   A synthetic flavoring agent that is a light yellow colored liquid of green leaf-wine odor. It is stable but may cause discoloration. It should be stored in glass or tin containers. It is used as flavoring for its wine note and has application in beverages and candy at 1 to 2 parts per million.

Allyl Caproate   See **Allyl *n*-Hexanoate**.

Allyl Cinnamate   A synthetic flavoring agent that is a fairly stable, hazy, colorless to light yellow colored liquid of cherry odor. It is used for its cherry note in flavors and has application in baked goods and candies at 1 to 2 parts per million.

Allyl-2,4-Hexadienoate   See **Allyl Sorbate**.

Allyl *n*-Hexanoate   A synthetic flavoring agent that is a colorless to yellow liquid with a strong pineapple-like odor. It is alkali and mineral acid unstable and should be stored in glass, tin, or resin-lined containers. It is used in flavors for its pineapple note and has applications in puddings, dessert gels, and candy at 10 to 30 parts per million. It is also termed allyl caproate.

Allyl Isothiocyanate   A synthetic flavoring agent that is a moderately stable, colorless to pale yellow liquid of pungent and irritating odor. It should be stored in glass containers. It is used as an artificial oil of mustard, and as an imitation horseradish flavor with application in condiments, meats, and pickles at 87 parts per million. It is also termed mustard oil.

Allyl Isovalerate   A synthetic flavoring agent that is a stable, colorless to light yellow liquid of fruity odor. It should be stored in glass or tin containers. It has usage in fruit flavors with applications in beverages, baked goods, ice cream, and candy at 8 to 50 parts per million.

Allyl Mercaptan   A synthetic flavoring agent that is a stable, colorless liquid of garlic-like odor. It should be stored in glass or tin containers. It is used in artificial garlic flavors for application in condiments at 3 parts per million, in baked goods at 2 parts per million. It is also termed 2 propylene-1 thiol.

Allyl Nonanoate   A synthetic flavoring agent that is a stable, colorless to light yellow liquid of fruity-cognac odor. It should be stored in glass or tin containers. It is used in fruit flavors like melon and pineapple for application in candy, ice cream, beverages at 0.70 to 5 parts per million.

Allyl *n*-Octanoate   A synthetic flavoring agent that is a colorless to light yellow liquid and has a fruity odor. It is alkali and mineral acid unstable and should be stored in glass, tin, and resin-lined containers. It is used to give flavors a fruity note and has application in dessert gels, puddings, beverages, and candy at 3 to 25 parts per million.

Allyl Phenylacetate   A synthetic flavoring agent that is a stable, colorless to light yellow liquid with a fruity odor of banana and honey. It should be stored in glass or tin containers. It is used in flavors for honey and has application in candy and baked goods at 10 to 15 parts per million.

Allyl Phenoxyacetate   A synthetic flavoring agent that is a stable, colorless to light yellow liquid of heavy fruit note odor. It should be stored in glass or tin containers. It is used in pineapple, quince, and fruit flavors with applications in candy and beverages at 1 to 3 parts per million.

Allyl Sorbate   A synthetic flavoring agent that is a colorless to light yellow liquid of sharp fruity odor. It is subject to polymerization and should be stored in glass or tin containers. It is used in pineapple and other fruit flavors which have application in puddings, candy, and beverages at 1 to 2 parts per million. It is also termed allyl-2, 4-hexadienoate.

Almond   A nut obtained from the almond tree *Prunus amygdalus*. It exists as a sweet or bitter variety, with the sweet being used as edible nuts and the bitter variety being used as a source of almond oil. The obtainable forms range from whole nuts to slices to powder. The nuts are used as snacks, as a garnish on pastries, and as a flavorant.

Almond Oil   The oil of the bitter almond after the removal of hydro-cyanic acid. It is a colorless to slightly yellow liquid having a strong principally almond-like aroma. It is used mainly in the pharmaceutical and cosmetic industry and also as a food flavoring agent.

Almond Paste   A paste made by cooking sweet and bitter almonds which have been ground and blanched in combination with sugar. It consists approximately of two parts almond to one part sugar. It is used in pastries and cakes.

Alpha-Tocopherol   See **Tocopherol**.

Alum   A preservative, the inclusive term for several aluminum-type compounds such as aluminum sulfate and aluminum potassium sulfate. It is used with EDTA to prevent discoloration of potatoes, and to maintain the firmness of shrimp packs. It is also used in pickles and pickle relish.

Aluminum Acetate   A general purpose food additive. A patent discusses its use with carboxymethylcellulose in stabilizing starch sponge food products.

Aluminum Ammonium Sulfate $AlH_4NO_8S_2$. A general purpose food additive that functions as a buffer and neutralizing agent. Its solubility is 1 g in 7 ml of water at 25°C and 1 g in 0.3 ml of boiling water. It is used in baking powders.

Aluminum Calcium Silicate An anticaking agent used in vanilla powder. It is also used in salt up to 2 percent.

Aluminum Caprate The aluminum salt of capric acid which is used as a binder, emulsifier, and anticaking agent.

Aluminum Caprylate The aluminum salt of caprylic acid which is used as a binder, emulsifier, and anticaking agent.

Aluminum Laurate The aluminum salt of lauric acid which is used as a binder, emulsifier, and anticaking agent.

Aluminum Myristate The aluminum salt of myristic acid which is used as a binder, emulsifier, and anticaking agent.

Aluminum Nicotinate. $C_{18}H_{12}AlNO_3O_6$. The aluminum salt of nicotinic acid. It is a source of niacin in foods of special dietary use.

Aluminum Oleate The aluminum salt of oleic acid which is used as a binder, emulsifier, and anticaking agent. It is practically insoluble in water.

Aluminum Oxide Al_2O_3. A dispersing agent used with or without silicon dioxide in mixtures for food or drug use which contain titanium dioxide. Its use level is up to 2 percent.

Aluminum Palmitate $C_{48}H_{93}AlO_6$. The aluminum salt of palmitic acid which is used as a binder, emulsifier, and anticaking agent.

Aluminum Sodium Sulfate $AlNaS_2O_8$. A general purpose food additive that functions as a buffer, neutralizing agent, and firming agent. It is anhydrous and slowly soluble in water. The dodecahydrate form is readily soluble in water. It is also termed soda alum.

Aluminum Stearate $C_{54}H_{105}AlO_6$. The aluminum salt of stearic acid which is used as a binder, emulsifier, and anticaking agent. It is used as a component of defoaming agents limited to use in processing beet sugar and yeast.

Aluminum Sulfate $Al_2(SO_4)_3$. A starch modifier and firming agent. The anhydrous form has a slow rate of solution while the hydrate form, $Al_2(SO_4)_3 \cdot 14H_2O$, has a solubility of 1 g in approximately 2 ml of water and a 1% solution pH of approximately 3.5. It is used up to 2 percent in combination with not more than 2 percent of 1-octenyl succinic anhydride. It is used as a firming agent in pickle and vegetable processing; as a processing aid in baked goods, gelatins, and puddings.

Amidated Pectin The low-methoxyl pectin that results when some of the methoxyl groups are transformed into amide groups during deesterification with ammonia. These pectins function best in applications between 30 and 65 percent soluble solids content and pH 3.0 to 4.5. They usually require no more calcium ions than are already present in the fruit to obtain gelation. The gels formed are heat-reversible. Applications include flans and tart glazing.

Amioca See **Waxy Maize Starch**.

Ammoniated Glycyrrhizin See **Glycyrrhizin**.

Ammonium Alginate $(C_6H_7O_6NH_4)_n$. A gum that is the ammonium salt of alginic acid. It is cold water soluble and forms viscous solutions. It functions as a stabilizer and thickener and its uses include bakery fillings, gravies, and toppings.

Ammonium Bicarbonate NH_4HCO_3. A dough strengthener and leavening agent. One gram dissolves in approximately 6 g of water. It reacts with heat to release carbon dioxide which causes dough to rise. It is used in baked goods such as cookies and baking powders. It is used mainly in products baked to a low moisture content; in high-moisture foods there is a retention of the ammonia odor.

Ammonium Carbonate A leavening agent and pH control agent that consists of ammonium bicarbonate and ammonium carbamate in varying proportions. One gram dissolves slowly in 4 ml of water. It is used in baked goods, baking powders, gelatins, puddings, and in the production of caramels.

Ammonium Caseinate The ammonium salt of casein. It has a high nutritional value and low sodium content and is used in foods and pharmaceuticals. See **Caseinates**.

Ammonium Chloride NH_4Cl. A dough conditioner and yeast food that exists as colorless crystals or white crystalline powder. Approximately 30 to 38 g dissolve in water at 25°C. The pH of a 1 percent solution at 25°C is 5.2. It is used as a dough strengthener and flavor enhancer in baked goods and as a nitrogen source for yeast fermentation. It is also used in condiments and relishes. Another term for the salt is ammonium muriate.

Ammonium Hydroxide NH_4OH. An alkaline that is a clear, colorless solution of ammonia which is used as a leavening, pH control agent, and as a surface finishing agent. It is used in baked goods, cheese, puddings, process fruits, and in the production of caramels.

Ammonium Muriate NH_4Cl. See **Ammonium Chloride**.

Ammonium Persulfate A bleaching agent for food starch that is used at up to 0.075 percent and with sulfur dioxide up to 0.05 percent.

Ammonium Phosphate Dibasic $(NH_4)_2HPO_4$. A general purpose food additive that is readily soluble in water, with approximately 57 g dissolving in 100 g of water at 0°C. A 1 percent solution has a pH of 7.6 to 8.2. It is used as a dough strengthener, firming agent, leavening agent, and pH control agent. Its uses include baked goods, alcoholic beverages, condiments, and puddings. In bakery products up to 0.25 part per 100 parts by weight of flour is used.

Ammonium Phosphate Monobasic $NH_4H_2PO_4$. A general purpose food additive which is readily soluble in water. A 1 percent solution has a pH of 4.3 to 5.0. It is used as a dough strengthener and leavening agent in baked goods and as a firming agent, and pH control agent in condiments and puddings. It is also used in baking powder with sodium bicarbonate and as a yeast food.

Ammonium Sulfate $(NH_4)_2SO_4$. A dough conditioner, firming agent, and processing aid which is readily soluble in water with a solubility of approximately 70 g in 100 g of water at 0°C. The pH of a 0.1 molar solution in water is approximately 5.5. It is used in caramel production and as a source of nitrogen for yeast fermentation. In bakery products, up to 0.25 part per 100 parts by weight of flour is used.

Ammonium Sulfite $(NH_4)_2SO_3$. An additive used in the production of caramel.

Amyl Acetate   A flavorant used in fruit and liquor flavors to provide a banana flavor characteristic.

Anhydrous Milkfat   See **Butter Oil**.

Anise   A spice that is the dried, ripe fruit of *Pimpinella anisum,* a small herb. The flavor is similar to fennel or licorice while the seed resembles caraway seed. It is used in beverages, soups, candy, liquors, and sweet pastries.

Anisyl Butyrate   A synthetic flavoring agent that is a stable, colorless liquid of sweet cassic odor. It should be stored in glass or tin containers. It will intensify vanilla flavor and is used as a fixative. It is used in ice cream, candy, and baked goods at 5 to 15 parts per million.

Anisyl Formate   A synthetic flavoring agent that is a fairly stable, colorless to light yellow liquid of floral odor. It should be stored in glass, tin, or resin-lined containers. It is used in berry flavors for applications in beverages, candy, and baked goods at 3 to 10 parts per million.

Anisyl Propionate   A synthetic flavoring agent that is a stable, colorless liquid with a heliotrope odor. It should be stored in glass or tin containers. It is used in small concentrations to intensify vanilla, plum, and quince flavor for applications in beverages, baked goods, and candy at 6 to 20 parts per million.

Annatto   A color source of yellowish to reddish-orange color obtained from the seed coating of the tree *Bixa orellanna.* The oil-soluble annatto consists mainly of bixin, a carotenoid soluble in fats and oils with the color which is produced being found in the fat portion of the food. It has a yellow hue, very good oxidation stability, fair light stability, and good heat stability but it is unstable above 125°C. The water-soluble annatto is norbixin (the product resulting when bixin is saponified and the methylethyl group is split off) which is dissolved as a potassium salt in lye. It is readily soluble in aqueous alkalis with the coloring occurring in the protein and starch fraction of the food. It has a yellow to orange hue and precipitates in most acid foods. The usage level is 0.5 to 10 parts per million in the finished food. It is used in sausage casings, oleomargarine, shortening, and cheese.

Annatto Extract   See **Bixin**.

Apple Vinegar   See **Cider Vinegar**.

Arabic   A gum obtained from breaks or wounds in the bark of *Acacia* trees. It dissolves in hot or cold water forming clear solutions which can be up to 50 percent gum acacia. The solubility in water increases with temperature. It is used in confectionary glazes to retard or prevent sugar crystallization and acts as an emulsifier to prevent fat from forming an oxidizable, greasy film. It functions as a flavor fixative in spray-drying to form a thin film around the flavor particle. It also functions as an emulsifier in flavor emulsions, as a cloud agent in beverages, and as a form stabilizer. It is also termed acacia.

Arabinogalactan   A gum, being the plant extract obtained from larch trees. It is soluble in hot and cold water, the water solutions up to 60 percent being fluid and above 60 percent forming a thick paste. It is stable over a wide pH range and is relatively unaffected by electrolytes. Its limited uses include dressings and pudding mixes. It is also termed larch gum.

Arginine   A nonessential amino acid that exists as white crystals or powder. It is soluble in water. It is used to improve the biological quality of the total protein in a food which contains naturally occurring primarily intact proteins and is used as a nutrient and dietary supplement.

Arrowroot   A starch obtained from *Maranta arundinacea*, a perennial that produces starchy rhizomes. It is neutral in flavor and of clear color. It is used as a thickener, using one-third to one-half as much as the flour or cornstarch level. It is used in fruit sauce, pie fillings, and puddings.

Artemisia   A flavoring derived from plants of the genus *Artemisia* which is used in alcoholic beverages. Its common name is wormwood.

Artificial Coloring   See **Colors category**.

Artificial Flavor   Any substance whose function it is to impart flavor and that is not derived from a spice, fruit, vegetable, edible yeast, bark, bud, herb, root, leaf, or other plant material, meat, seafood, eggs, poultry, dairy, or fermentation products. See **Flavors category**.

Ascorbic Acid $C_6H_8O_6$. It is termed vitamin C, a water-soluble vitamin that prevents scurvy, helps maintain the body's resistance to infection, and is essential for healthy bones and teeth. It is the most easily destroyed vitamin and processing is recommended in stainless steel or glass. Storage at below $-18°C$ is recommended. In its dry form it is nonreactive, but in solution it readily reacts with atmospheric oxygen and other oxidizing agents. One part ascorbic acid is equivalent to one part erythorbic acid. It is used as a vitamin supplement in beverages, potato flakes, and breakfast foods; and as a dough conditioning agent to strengthen and condition bread roll doughs. It is also used as an antioxidant to increase shelf life in canned and frozen processed foods. It is used in conjunction with BHA, BHT, and propyl gallate to regenerate them following the chemical changes they undergo when they prevent fat rancidity in bologna and other meats. Other forms of ascorbic acid are isoascorbic (erythorbic) acid, sodium ascorbate, and sodium isoascorbate.

Ascorbyl Palmitate $C_{22}H_{38}O_7$. An antioxidant formed by combining ascorbic acid with palmitic acid. Ascorbic acid is not fat soluble but ascorbyl palmitate is, thus combining them produces a fat-soluble antioxidant. It exists as a white or yellowish white powder of citric-like odor. It is used as a preservative for natural oils, edible oils, colors, and other substances. It acts synergistically with alpha-tocopherol in oils/fats. It is used in peanut oil at a maximum level of 200 mg/kg individually or in combination.

Ascorbyl Stearate   An antioxidant used in peanut oil at a maximum use level of 200 mg/kg individually or in combination.

Aspartame $C_{14}H_{18}N_2O_5$. A synthetic sweetener that is a dipeptide, providing 4 calories per gram. It is synthesized by combining the methyl ester of phenylalanine with aspartic acid, forming the compound N-L-alpha-aspartyl-L-phenylalanine-1-methyl ester. It is approximately 200 times as sweet as sucrose and tastes similar to sugar. Its minimum solubility is at pH 5.2, its isoelectric point. Its maximum solubility is at pH 2.2. The solubility increases with temperature. Aspartame has a certain instability in liquid systems which results in a decrease in sweetness. It decomposes to aspartylphenylalanine or to diketopiperazine (DKP) and neither of these forms is sweet. The stability of aspartame is a function of time, temperature, pH, and water activity. It is not usually used in baked goods because it breaks down at the high baking temperatures. It contains phenylalanine, which restricts its use for those afflicted with phenylketonuria, the inability to metabolize phenylalanine. Approved applications include cold breakfast cereals, desserts, topping mixes, chewing gum, powdered beverages, and frozen desserts. The usage level ranges from 0.01 to 0.02 percent. It is also approved for use in soft drinks, where it is used as a replacement for saccharin or as a admixture to saccharin-containing diet soft drinks.

Aspartic Acid $C_4H_7NO_4$. A nonessential amino acid that exists as colorless or white crystals of acid taste. It is slightly soluble in water. It functions to improve the biological quality of a total protein in a food containing naturally occurring primarily intact protein and as a nutrient and dietary supplement.

Azodicarbonamide $C_2H_4N_4O_2$. A dough conditioner that exists as a yellow to orange-red crystalline powder practically insoluble in water. It is used in aging and bleaching cereal flour to produce a more manageable dough and a lighter, more voluminous loaf of bread. It is used in bread flours and bread as a dough conditioner. It can be used with the oxidizing agent potassium bromate. A typical use level is less than 45 parts per million.

B

Babassu Oil The oil obtained from the nut of the babassu palm, which is native to Brazil. It is similar to coconut oil and acts as a substitute, being used in vegetable fat-based products.

Baking Powder A leavening agent that consists of a mixture of sodium bicarbonate, one or more leavening agents such as sodium aluminum phosphate or monocalcium phosphate, and an inert material such as starch. The inert material keeps the reactive components physically separated and minimizes premature reaction. It should yield not less than 12 percent of available carbon dioxide.

Baking Soda Bicarbonate of soda, chemically known as sodium bicarbonate. It liberates carbon dioxide, a leavening gas, when heated or mixed with an acid. It is a component of baking powder and is used as a leavening agent.

Bakers' Yeast Glycan The dried cell walls of yeast, *Saccharomyces cerevisiae*, obtained from brewing. Bakers' yeast glycan is used as an emulsifier and thickener in salad dressing.

Barley A cereal grain of which there are winter and spring types. It is used in malting (the conversion of grain to malt used in beer production) as malted barley. Malt flour is used in baking, cereals, and sauces. Pearled barley, in which the hull and outer kernel are removed by abrasive action, is found in barley soups. Barley flour and flakes are used in baked products. Barley is high in carbohydrates and contains protein, calcium, phosphorus, and B vitamins.

Barley, Malted   See **Malted Barley**.

Basil (Sweet Basil)   A spice obtained from the dried leaves and tender stems of *Ocimum basilicum L.* The fresh basil resembles licorice in flavor and the dried leaves have a lemony anise-like quality. This delicate herb can be used generously and has an affinity for tomato-based products. It is used in tomato base recipes, with vegetables, and in tomato sauces.

Bay Leaves   A spice flavoring that consists of the dried leaves obtained from the evergreen tree *Laurus nobilis*, also called sweet bay or laurel tree. They have a sweet, herbaceous flavor and are used as an herb that is aromatic when crushed and find use in meat, soup, and stew.

Beeswax   The purified wax obtained from the honeycomb of the bee is insoluble in water and is sparingly insoluble in cold alcohol. It is used to glaze candy and is used in chewing gum, confections, and as a flavoring agent.

Beet Extract   A natural red colorant obtained from beets that is of very good water solubility and has fair pH stability, poor heat stability, and good light stability. It is colored by betacyanins which include red and yellow compounds, the major red pigment being betanin. The betanin accounts for 75 to 95 percent of the total pigment content. It is available in concentrate and powder form and is used in yogurt, beverages, candies, and desserts.

Beets, Dehydrated   See **Beet Extract**.

Beet Sugar   See **Sugar**.

Benne   See **Sesame Seed**.

Bentonite   A general purpose additive that is used as a pigment and colorant and to clarify and stabilize wine.

Benzaldehyde C_7H_6O. A flavorant that is the artificial oil of almond, occurring in bitter almond kernels. It is a colorless liquid having an odor resembling that of bitter almond oil and a burning taste. It oxidizes in air to benzoic acid. It provides a nutty almond flavor. It is also termed benzoic aldehyde.

Benzoate of Soda   See **Benzoic Acid**.

Benzoic Acid $C_7H_6O_2$. A preservative that occurs naturally in some foods such as cranberries, prunes, and cinnamon. It is most often used in the form of the sodium benzoate because of the low aqueous solubility of the free acid. Sodium benzoate is 180 times as soluble in water at 25°C. The salt in solution is converted to the acid which is the active form. The optimum pH range for microbial inhibition is pH 2.5 to 4.0. It is used in acid foods such as carbonated beverages, fruit juice, and pickles. It is also termed benzoate of soda.

Benzoic Aldehyde   See **Benzaldehyde**.

Benzoyl Peroxide $C_{14}H_{10}O_4$. A colorless, crystalline solid with a faint odor of benzaldehyde resulting from the interaction of benzoyl chloride and a cooled sodium peroxide solution. It is insoluble in water. It is used in specified cheeses at 0.0002 percent of milk level. It is used for the bleaching of flour, slowly decomposing to exert its full bleaching action, which results in whiter flour and bread.

Beta-Apo-8′-Carotenal $C_{30}H_{40}O$. A colorant that is a carotenoid producing a light to dark orange hue. It has fair light stability, poor oxidation stability, and good pH stability. It is insoluble in water but is available in water-dispersible, oil-dispersible, and oil-soluble forms. It has vitamin A activity. It is used in cheese, and cheese sauces, and dressings. The maximum usage level is 33 parts per million. Related colorants are beta-carotene and canthaxanthin.

Beta-Carotene   A colorant that is a carotenoid producing a yellow to orange hue. It has good tinctorial strength, fair light stability, poor oxidation stability, and good pH stability. It is insoluble in water but is available as water-dispersible, oil-dispersible, and oil-soluble forms. It has vitamin A activity. It has a natural resistance to ascorbic acid reduction in beverages and thus is used in orange-colored liquid products. It is used in margarine, oils, cheese, and puddings at levels required to produce the desired color. Related colorants are canthaxanthin and beta-apo-8′-carotenal.

BHA   See **Butylated Hydroxyanisole**.

BHT   See **Butylated Hydroxytoluene**.

Bicarbonate of Soda   See **Sodium Bicarbonate**.

Biotin $C_{10}H_{16}N_2O_3S$. A water-soluble vitamin that is a nutrient and dietary supplement. It is relatively stable to heat and storage and is found in eggs, liver, peanuts, milk, and meat. It functions in the metabolism of carbohydrates, proteins, and fats. It is essential for the activity of many enzyme systems.

Biphenyl   A flavoring agent, the maximum usage level being 2 parts per million in beverages, candy, frozen desserts, and baked goods.

Birch   An artificial flavoring used in soft drinks such as birch beer.

Bixin $C_{25}H_{30}O_4$. A carotenoid that is the main coloring component of annatto. It is obtained from the *Bixa orellana* tree. Bixin is soluble in fats and oils and the produced color is found in the fat fraction of the food. It has a yellow hue, very good oxidation stability, fair light stability, good heat stability but is poor at very high temperatures, such as above 125°C. One part bixin is equivalent to 1.5 parts carotene. It is used at 0.5 to 10 parts per million in finished foods, such as margarine, salad dressings, popcorn oil and baked goods. It is also termed annatto extract. See **Annatto**.

Bleached Flour   Flour that has been whitened by the removal of the yellow pigment. The bleaching can be obtained during the natural aging of the flour or can be accelerated by chemicals that are usually oxidizing agents which oxidize the carotenoid pigments to a nearly colorless product. The oxidizing agents also improve the flour performance by their effect on the protein. The process improves the baking quality by allowing the formation of high ratio cakes that would be likely to collapse if prepared with untreated flour.

Bran   The seed husk or outer coatings of cereals, such as wheat, rye, and oats that is separated from the flour. It is used in bran flakes cereal.

Bread Flour   A hard-wheat flour, which generally contains in excess of 10.5 percent protein and is obtained from straight or long patent flours. These flours have high absorption and good mixing tolerance which make them suitable in yeast-leavened breads.

Brilliant Blue FCF   See **FD&C Blue #1**.

Bromated Flour   A white flour to which potassium bromate is added at a level not to exceed 50 parts per million. It is used in baked goods.

Bromelin   A proteolytic enzyme obtained from pineapple which functions to break down the protein. It is used in tenderizing beef and in chill-proofing beer to reduce haze formation. It also has some use in bread baking for the reduction of mixing time and dough viscosity. It is present in meat tenderizers.

Brominated Vegetable Oil   **(BVO)** A vegetable oil whose density has been increased to that of water by being combined with bromine. Flavoring oils are dissolved in the brominated oil which can then be added to fruit drinks. The adjustment of the specific gravity makes it possible to obtain stable finished beverages; otherwise, if the oil phase gravity is too low the emulsion will form a ring and if it is too high a white precipitate may form. It is also used in formulating cloud agents. Its use is limited to 15 parts per million.

Brown Sugar   A sweetener that consists of sucrose crystals covered with a film of cane molasses which gives it the characteristic color and flavor. There are three grades: light, medium, and dark which vary in sucrose content and color. It is used in baked goods, glazes, toppings, and fillings.

Buckwheat   A member of the grass family that is usually marketed as a cereal grain, buckwheat grain. It is high in carbohydrates and is an excellent source of protein. Buckwheat flour is used in pancake mixes and crepe mixes. Buckwheat groats are the kernels that remain after removal of the hulls and can be used in breakfast cereals, stuffings, and health foods.

Bulgur   A precooked cracked wheat that retains the bran and germ fraction of the grain. It nutritionally resembles whole wheat and is sometimes termed parboiled wheat. It is an excellent source of whole grain, protein, and carbohydrates. It is reconstituted by cooking or soaking in liquid. It can be used in bread, casseroles, salads, or can be eaten as such.

Butter   A source of milk fat obtained from cream by a churning process which results in the production of butter fat and buttermilk. Butter is 80 percent fat, melts completely at about 38°C, and contains more than 40 percent solid fat at 10°C. It is graded according to flavor, body, color, and salt into the following groups: U.S. grade AA, U.S. grade A, U.S. grade B, and U.S. grade C.

Butter, Clarified   Butter that has undergone purification by the removal of solid particles or impurities that may affect the color, odor, or taste.

Butter Fat.   See **Milkfat**.

Buttermilk   The product that remains when fat is removed from milk or cream in the process of churning into butter. Cultured buttermilk is prepared by souring buttermilk, or more commonly skim milk, with a suitable culture that produces a desirable taste and aroma. It is used as a beverage, as an ingredient in baked goods, and in dressings.

Buttermilk, Dried   The powder form of buttermilk. It is similar in composition to nonfat dry milk but of higher fat concentration, much of which is phospholipids which provide good emulsifying and whipping properties. It is used in dry mix, desserts, soups, and sauces.

Butter Oil   The clarified fat portion of milk, cream, or butter obtained by the removal of the nonfat constituent. It contains not less than 99.7 percent milk fat, not more than 0.2 percent moisture, and not more than 0.05 percent milk solids nonfat. It is used in frozen desserts, puddings, and syrups. It is also termed anhydrous milkfat, or ghee.

Butyl Butyryllactate $C_{11}H_{20}O_4$. A synthetic flavoring agent that is a stable, colorless to light yellow liquid with the odor of cooked butter. It is miscible with alcohol and most fixed oils, soluble in propylene glycol, and insoluble in glycerine and water. It should be stored in glass, tin, or resin-lined containers. It is used in butter flavors with applications in baked goods, and candy at 14 to 60 parts per million.

Butyl Heptanoate   A synthetic flavoring agent that is a stable, colorless liquid of fruity odor. It is stored in glass or tin containers. It is used in flavors such as apple, blackberry, and ginger beer with applications in candy, and baked goods at 25 parts per million.

Butylated Hydroxyanisole (BHA) $C_{11}H_{16}O_2$. An antioxidant that imparts stability to fats and oils and should be added before oxidation has started. It is a mixture of 3-tert-butyl-4-hydroxyanisole and 2-tert-butyl-4-hydroxyanisole. In direct addition, the fat or oil is heated to 60 to 70°C and the BHA is added slowly under vigorous agitation. The maximum concentration is 0.02 percent based on the weight of the fat or oil. It may protect the fat-soluble vitamins A, D, and E. It is used singly or in combination with other antioxidants. It is used in cereals, edible fat, vegetable oil, confectionary products, and rice.

Butylated Hydroxytoluene (BHT) $C_{15}H_{24}O$. An antioxidant that functions similarly to butylated hydroxyanisole (BHA) but is less stable at high temperatures. It is also designated 2,6-di-tert-butyl-para-cresol. See **Butylated Hydroxyanisole**.

1,3-Butylene Glycol $C_4H_{10}O_2$. A humectant that is a colorless, viscous liquid. It is miscible with water, acetone, and ether but is immiscible with fixed oils. It dissolves most essential oils and synthetic flavoring substances.

Butylhydroquinone See **Tertiary Butylhydroquinone**.

Butylparaben See **Parabens**.

Butyric Acid $C_4H_8O_2$. A fatty acid that is commonly obtained from butter fat. It has an objectionable odor which limits its use as a food acidulant or antimycotic. It is an important chemical reactant in the manufacture of synthetic flavoring, shortening, and other edible food additives. In butter fat, the liberation of butyric acid which occurs during hydrolytic rancidity makes the butter fat unusable. It is used in soy milk type drinks and candies.

Cacao Butter   See **Cocoa Butter**.

Caffeine $C_8H_{10}N_4O_2$. A white powder or needles that are odorless and have a bitter taste, occurring naturally in tea leaves, coffee, cocoa, and cola nuts. It is a food additive used in soft drinks for its mildly stimulating effect and distinctive taste note. It is used in cola-type beverages and optional in other soft drinks up to 0.02 percent.

Cake Flour   A soft wheat flour that is generally a short patented flour containing less than 10 percent protein. Such flours are low in water absorption and are of short mixing time and tolerance. It is used in chemically leavened cakes, cookies, and pastries.

Calciferol $C_{28}H_{44}O$. A fat-soluble vitamin, termed vitamin D_2, which is stable unless oxidized. It is necessary for growth and maintenance of teeth and bones and the normal utilization of calcium and phosphorus; it is used medicinally in the treatment of rickets and as a dietary supplement. Its sources include fish, liver, and vitamin D fortified milk.

Calcium   Ca. An alkaline earth element that contributes toward bone and teeth formation, muscle contraction, and blood clotting. It occurs in milk, vegetables, and egg yolk.

Calcium Acetate $Ca(C_2H_3O_2)_2$. The calcium salt of acetic acid which functions as a sequestrant and mold control agent. It contains approximately 25 percent calcium. It is a white odorless powder which is readily soluble in water with a solubility of approximately 37 g in 100 g water at 0°C. Its solubility decreases with increasing temperature, with a solubility of approximately 29 g in 100 g of H_2O at 100°C.

Calcium Acid Phosphate   See **Monocalcium Phosphate**.

Calcium Alginate   The calcium salt of alginic acid which functions as a stabilizer and thickener. The partial obtainment of calcium alginate by the reaction of the water-soluble sodium alginate with calcium ions is used to obtain viscosity and gel formation. It is used in icings, imitation pulp, dessert gels, and fabricated fruits.

Calcium Aluminum Silicate   An anticaking agent that permits the free flow of dry ingredients. It is used in salt and dry mixes.

Calcium Ascorbate $C_{12}H_{14}CaO_{12} \cdot 2H_2O$. The salt of ascorbic acid which is a white to slightly yellow crystalline powder. It is soluble in water and the pH of a 10 percent solution is 6.8 to 7.4. It functions as an antioxidant and preservative. See **Ascorbic Acid**.

Calcium Biphosphate   See **Monocalcium Phosphate**.

Calcium Bromate $Ca(BrO_3)_2 \cdot H_2O$. A dough conditioner, maturing and bleaching agent which exists as a white crystalline powder. It is very soluble in water and is used in flour and dough.

Calcium Caprate   The calcium salt of capric acid. It is used as a binder, emulsifier, anticaking agent, and as a general additive.

Calcium Caprylate   The calcium salt of caprylic acid. It is used as a binder, emulsifier, anticaking agent, and as a general additive.

Calcium Carbonate $CaCO_3$. The calcium salt of carbonic acid which is used as an anticaking agent and dough strengthener. It is available in varying particle sizes ranging from coarse to fine powder. It is practically insoluble in water and alcohol, but the presence of any ammonium salt or carbon dioxide increases its solubility while the presence of any alkali hydroxide reduces its solubility. It has a pH of 9 to 9.5. It is the primary source of lime (calcium oxide) which is made by heating limestone in a furnace. Calcium carbonate is used as a filler in baking powder, for calcium enrichment, and as a mild buffering agent in doughs, in dry mix desserts as a source of calcium ions, and as a neutralizer in antacids. It is also termed limestone.

Calcium Carrageenan   See **Carrageenan**.

Calcium Caseinate   The calcium salt of casein. Properties include low viscosity, settling out of water, opaqueness, no heat stability, and chalky texture. It contains the range of essential amino acids present in sodium caseinate but has a higher concentration of calcium. It is useful in applications requiring low absorption properties. It is used as a protein source in imitation cheese, and in special diet foods to replace sodium caseinate where sodium must be restricted. It is used to improve the whipping properties of vegetable whipped toppings, and as a binder. See **Caseinates**.

Calcium Chloride $CaCl_2$. A general purpose food additive, the anhydrous form being readily soluble in water with a solubility of 59 g in 100 ml of water at $0°C$. It dissolves with the liberation of heat. It also exists as calcium chloride dihydrate ($CaCl_2 \cdot 2H_2O$), being very soluble in water with a solubility of 97 g in 100 ml at $0°C$. It is used as a firming agent for canned tomatoes, potatoes, and apple slices. In evaporated milk, it is used to adjust the salt balance so as to prevent coagulation of milk during sterilization at levels not more than 0.1 percent. It is used with disodium EDTA to protect the flavor in pickles and as a source of calcium ions for reaction with alginates to form gels.

Calcium Citrate $Ca_3(C_6H_5O_7)_2 \cdot 4H_2O$. The calcium salt of citric acid which functions as a sequestrant, buffer, and firming agent. It is a white, odorless powder which is slightly soluble in water. It is used as a firming agent for peppers and lima beans, and is used to improve the baking properties of flour.

Calcium Diacetate   The salt of acetic acid which is used as a preservative and sequestrant.

Calcium Disodium EDTA   See **EDTA**.

Calcium Gluconate $C_{12}H_{22}CaO_{14}$. A white crystalline granule or powder that functions as a firming agent, formulation aid, sequestrant, and stabilizer. At room temperature the anhydrous form has a solubility of approximately 1 g in 30 ml of water, which improves in boiling water to approximately 1 g in 5 ml of water. It also exists as calcium gluconate (monohydrate) of $C_{12}H_{22}CaO_{14} \cdot H_2O$ formulation. It is used as a source of calcium ions for sodium alginate gels, and as a calcium fortifier in baked goods, puddings, and dairy product analogs. It functions as a coagulation aid in milk and instant pudding powders, and as a means of masking the bitter aftertaste of some artificial sweeteners.

Calcium Glycerophosphate $C_3H_7CaO_6P$. A nutrient and dietary supplement which is a white odorless powder of poor water solubility. It is used in dental impression material and baking powder.

Calcium Hexametaphosphate See **Calcium Metaphosphate**.

Calcium Hydrate See **Calcium Hydroxide**.

Calcium Hydroxide $Ca(OH)_2$. A general food additive made by adding water to calcium oxide (lime). It has poor water solubility with a solubility of 0.185 g in 100 g water at 0°C. The pH of a water solution at 25°C is approximately 12.4. It is used to promote dispersion of ingredients in sauces, creamed spinach, and a frozen pea/potato dish. It is used at 0.1 percent to stabilize the potassium iodide of iodized salt, and it used to be used as a neutralizer for soured cream prior to buttermaking. It is also termed hydrated lime, calcium hydrate, and slaked lime.

Calcium Hydroxyphosphate A food additive used in table salt.

Calcium Iodate CaI_2O_6. A source of iodine that is a white powder of slight solubility in water, but more soluble in water containing iodides or amino acids. It is more stable than the iodide form. It is used as a dough conditioner in bread and is a source of iodine in table salts.

Calcium Lactate $C_6H_{10}CaO_6$. The calcium salt of lactic acid which is soluble in water with a solubility of 3.4 g per 100 g of water at 20°C and is very soluble in hot water. It is available as monohydrate, trihydrate, and pentahydrate. The trihydrate and pentahydrate have solubilities of 9 g in 100 ml H_2O at 25°C. It contains approximately 14 percent calcium. It is used in the stability and texture improvement of canned fruits and vegetables by the conversion of the labile pectin to the less soluble calcium pectate and thereby prevents structural collapse during cooking. It is used in angel food cake, whipped toppings, and meringues to increase protein extensibility which results in an increase of foam volume. It is also used in calcium fortified foods such as infant foods and is used to improve the properties of dry milk powder.

Calcium Lactobionate $C_{24}H_{42}CaO_{24}$. The calcium salt of lactobionic acid which is produced by the oxidation of lactose. It is soluble in water and is used as a firming agent in dry pudding mixes.

Calcium Laurate The calcium salt of lauric acid which is used as a binder, emulsifier, and anticaking agent.

Calcium Metaphosphate A sequestrant that is also termed calcium hexametaphosphate.

Calcium Myristic The calcium salt of myristic acid which is used as a binder, emulsifier, and anticaking agent.

Calcium Oleate The calcium salt of oleic acid which is used as a binder, emulsifier, and anticaking agent.

Calcium Oxide CaO. A general food additive consisting of white granules or powder of poor water solubility. It is obtained by heating limestone (calcium carbonate) in a furnace. It is also termed lime or quicklime. It is used as an anticaking agent, firming agent, and nutritive supplement in applications such as grain products and soft candy.

Calcium Palmitate The calcium salt of palmitic acid which is used as a binder, emulsifier, and anticaking agent.

Calcium Pantothenate $C_{18}H_{32}CaN_2O_{10}$. A nutrient and dietary supplement which is the calcium chloride double salt of calcium pantothenate. It is a white powder of bitter taste and has a solubility of 1 g in 3 ml of water. It is used in special dietary foods.

Calcium Pectinate The salt of pectin which is obtained from citrus or apple fruit. It results from the interaction of low-methoxyl pectin with calcium ions to form a gel. It is used as a gel coating for meat products and to form food gels. See **Low-Methoxyl Pectin**.

Calcium Peroxide CaO_2. A dough conditioner which exists as a white or yellowish powder or granule that is insoluble in water. It improves dough strength, grain, and texture, and increases absorption and crumb resiliency. It is used in bakery products.

Calcium Phosphate A compound existing in several forms which include the monobasic, dibasic, and tribasic forms of calcium phosphate. As calcium phosphate monobasic, also termed monocalcium phosphate, calcium biphosphate, acid calcium phosphate, it is used as a leavening agent and acidulant. Calcium phosphate dibasic, also termed dicalcium phosphate dihydrate, is used as a dough conditioner and mineral supplement. Calcium phosphate tribasic, also termed tricalcium phosphate, precipitated calcium phosphate, is used as an anticaking agent, mineral supplement, and conditioning agent.

Calcium Phosphate, Dibasic Anhydrous See **Dicalcium Phosphate, Anhydrous**.

Calcium Phosphate, Dibasic Hydrous See **Dicalcium Phosphate, Dihydrate**.

Calcium Phosphate, Monobasic See **Monocalcium Phosphate**.

Calcium Phosphate Tribasic See **Tricalcium Phosphate**.

Calcium Phytate A chelating agent that is used to bind metallic impurities in foods to prevent discoloration and off-flavor.

Calcium Propionate $C_6H_{10}CaO_4$. The salt of propionic acid which functions as a preservative. It is effective against mold, has limited activity against bacteria, and no activity against yeast. It is soluble in water with a solubility of 49 g per 100 ml of water at 0°C and insoluble in alcohol. It is less soluble than sodium propionate. Its optimum effectiveness is up to pH 5.0 and it has reduced action above pH 6.0. It is used in bakery products, breads, and pizza crust to protect against mold and "rope." It is also used in cold-pack cheese food, and pie fillings. Typical usage level is 0.2 to 0.3 percent and 0.1 to 0.4 percent based on flour weight.

Calcium Pyrophosphate $Ca_2P_2O_7$. A nutrient and dietary supplement that exists as a white odorless powder, insoluble in water. It is used in dental impression materials and as a buffer.

Calcium Saccharate A compound of calcium hydroxide and sugar that is used to improve the whipping ability of whipping cream, and to standardize the acidity in dairy products.

Calcium Saccharin $C_{14}H_8CaN_2O_6S_2 \cdot 3\frac{1}{2}$ H_2O. A sweetener that is the calcium form of saccharin, existing as white crystals or powder with a solubility of 1 g in 1.5 ml of water. Sodium saccharin is the more common form, but calcium saccharin is available for nonsodium diets. In this form it is about 500 times as sweet as sucrose. See **Saccharin**.

Calcium Silicate An anticaking agent that exists in different forms of which the most common are $CaSiO_3$, Ca_2SiO_4, and Ca_3SiO_5, which are insoluble in water. It is used in salt to enhance flowability under extremely high humidity conditions. It is also used in baking powder and fabricated chips to absorb water or other liquids.

Calcium Sorbate   A preservative that is the calcium salt of sorbic acid but is not the most common form. Its solubility in water or fat is very limited and therefore it is used on surfaces for preservation. It is permitted in cheese and wrapping materials.

Calcium Stearate   The calcium salt of stearic acid which functions as an anticaking agent, binder, and emulsifier. It is used in garlic salt, dry molasses, vanilla and vanilla-vanillin powder, salad dressing mix, and meat tenderizer. It can be used for mold release in the tableting of pressed candies.

Calcium Stearoyl Lactylate   A dough conditioner, whipping agent, and emulsifier that is the calcium salt of the reaction product of lactic acid and stearic acid, neutralized to the calcium salt. The sister product is sodium stearoyl-2-lactylate. It is used to increase dough tolerance to mixing, to increase the gas retention properties of dough which reduces the proofing time and increases the loaf volume. It is used as a whipping agent in liquid and frozen egg white and vegetable-type toppings. It is also used in dehydrated potatoes as a conditioning agent.

Calcium Sulfate   A general additive available as calcium sulfate anhydrous, $CaSO_4$, made by the high-temperature calcining of gypsum which is then ground and separated and as calcium sulfate dihydrate, $CaSO_4 \cdot 2H_2O$, which is made by grinding and separating gypsum containing about 20 percent water of crystallization. Calcium sulfate anhydrous contains approximately 29 percent calcium, and calcium sulfate dihydrate contains approximately 23 percent calcium. It is used, among other things, as a filler and baking powder for standardization purposes; a firming agent in canned potatoes, tomatoes, carrots, lima beans, and peppers; in dough as a source of calcium ions because the absence of calcium ions causes bread dough to be soft and sticky and to produce bread of poor quality; in soft-serve ice cream to produce dryness and stiffness; as a calcium ion source for reaction with alginates to form dessert gels; and as a calcium source for food enrichment.

Cane Sugar   See **Sugar**.

Canthaxanthin $C_{40}H_{52}O_2$. A synthetic red colorant that is the carotenoid of most intense red color. It is available in an oil-soluble, oil-dispersible, and water-dispersible forms. It has fair pH, heat, light, and chemical stability with a low tinctorial strength. Unlike the carotenoids beta-carotene and beta-apo-8′-carotenal, it does not possess vitamin A activity. Maximum usage level is 66 parts per million. Uses include carbonated soft drinks, salad dressing, and spaghetti sauce.

Caramel A colorant that is an amorphous, dark-brown product resulting from the controlled heat treatment of carbohydrates such as dextrose, sucrose, and malt syrup. It is available in liquid and powdered forms, providing shades of brown. In coloring a food with caramel, the food components must have the same charge as the particles of caramel, otherwise the particles will attract one another and precipitate out. Caramel can exist as several types, for example, acid-proof caramel of negative charge which is used in carbonated beverages; acidified solutions, bakers' and confectioners' caramel which are used in baked goods; and dried caramel for dry mixes. A major use is in coloring beverages such as colas and root beers and in baked goods.

Caraway A spice that is a seed obtained from the tree *Carum carvi.* It has a flavor similar to dill. It is used in rolls, bread, meats, and some cheeses.

Carbon Dioxide CO_2. A gas obtained during fermentation of glucose (grain sugar) to ethyl alcohol. It is used in pressure-packed foods as a propellant or aerating agent and is used in the carbonation of beverages. It is released as a result of the acid carbonate reaction of leavening agents in baked goods to produce an increase in volume. As a solid, it is termed dry ice and is used for freezing and chilling.

Carbonated Water A beverage made by absorbing carbon dioxide in water. The carbon dioxide influences flavor because increased carbonation increases mouth feel. Gas retention is more common in low-calorie-type beverages because of the absence of sugar solids.

Carboxymethylcellulose (CMC) A gum that is a water-soluble cellulose ether manufactured by reacting sodium monochloroacetate with alkali cellulose to form sodium carboxymethylcellulose. It dissolves in hot or cold water and is fairly stable over a pH range of 5.0 to 10.0, but acidification below pH 5.0 will reduce the viscosity and stability except in a special acid-stable type of CMC. A variety of types are available which differ in viscosity and degree of substitution (the number of sodium groups per unit). It functions as a thickener, stabilizer, binder, film former, and suspending agent. It is used in a variety of foods to include dressings, ice cream, baked goods, puddings, and sauces. The usage range is from 0.05 to 0.5 percent.

Cardamon A spice that is a dried, ripe seed of *Elettaria cardamomum*, a biennial plant. It has a pungent aroma and is reddish-brown in color. The flavor is sweet and spicy with a camphoraceous note. It is used in whole form to flavor hot fruit punches, pickles, and marinades. It is used in the ground form in bread, cookies, desserts, and meats.

Carmine The red colorant aluminum lake of carminic acid which is the coloring pigment obtained from dried bodies of the female insect *Coccus cacti*. It is brilliant red to purplish in color, having a low tinctorial strength, and can be solubilized in ammonia. It is used in a pink color in coatings.

Carnauba Wax A general purpose food additive that is a hard and brittle wax. It is obtained from leaf buds and leaves of the Brazilian wax palm *Copernicia cerifera*. It is the hardest wax known and is used in candy glaze.

Carob A cocoa substitute obtained from the pods of the carob tree *Ceratonia Siliqua*. The pods are kibbled, roasted, and ground into a powder which is similar in appearance and fragrance to cocoa powder. Carob powder has less than 1 percent fat and 42 to 48 percent sugar, while cocoa powder has approximately 23 percent fat and 5 percent sugar. Cocoa does not contain any measurable amounts of fructose so the presence of carob in cocoa can be detected by the presence of fructose. It is used in candy, drinks, bakery products, and dairy applications, and as a single ingredient in health food products.

Carob Gum See **Locust Bean Gum**.

Carotene $C_{40}H_{56}$. A colorant and provitamin, being a hydrocarbon which is a subgroup of the carotenoids (yellow, orange, or red pigments) while the other subgroup is xanthophylls. It functions as a colorant with beta-apo-8'-carotenal being a red-orange carotenoid and beta-carotene being a yellow carotenoid. It is also a vitamin A precursor that is converted by the body to vitamin A. It is used in ice cream, cheese, and other dairy products.

Carrageenan A gum that is a seaweed extract obtained from red seaweed *Chondrus crispus* or *Gigartina mammillosa*. It exists as various salts or mixed salts of a sulfate ester. It is classified mainly as kappa, iota, and lambda types which differ in solubility and gelling properties. The kappa and iota types require hot water (above 71°C) for complete solubility and can form thermally reversible gels in the presence of potassium and calcium cations, respectively. The kappa gels are brittle with syneresis while the iota gels are more elastic without syneresis. The lambda type is cold-water soluble and does not form gels. Kappa and iota carrageenan are very reactive with milk protein and are used with other gums to prevent whey-off in dairy products. Carrageenan is used to stabilize milk protein at 0.01 to 0.03 percent and to form water gels at 0.5 to 1.0 percent. Its uses include dairy products, water gel desserts, and low-calorie jellies. A typical use level in water systems is 0.2 to 1.0 percent and in milk systems is 0.01 to 0.25 percent.

Carubin See **Locust Bean Gum**.

Casein The principal milk protein which is prepared commercially from skim milk by the precipitation with lactic, hydrochloric, or sulfuric acid. It can also be produced by the use of lactic acid-producing bacteria. Caseins are usually identified according to the acid used, such as lactic acid casein, hydrochloric acid casein, and sulfuric acid casein. The principal form in which casein is used is as casein salts, of which sodium and calcium caseinate are the most common. Rennet casein is obtained from skim milk by the precipitation with a rennet-type enzyme. Casein is used in the protein fortification of cereals and bread, and in fabricated cheeses.

Caseinates Salts of casein that are produced by neutralizing acid casein to pH 6.7 with calcium or sodium hydroxide, producing the most common forms which are calcium caseinate or sodium caseinate. Other forms of casein are potassium, and ammonium caseinate. The caseinates provide a source of protein and function as emulsifiers, water binders, and whipping aids. The relative water absorption of casein salts is: calcium caseinate — 130 percent, potassium caseinate —200 percent, sodium caseinate — 250 percent. Its uses include processed meats, whipped toppings, coffee whiteners, egg substitutes, and diet foods.

Cayenne Pepper See **Pepper, Cayenne**.

Celery Seed A spice made from the dried, ripe fruit of the herb *Apium* *graveolens*, related to the parsley family. It is used lightly so as not to dominate in flavor. It is used in sauces, salads, meats, and soups.

Cellulose $(C_6H_{10}O_5)_n$. A carbohydrate polymer made up of glucose units. It consists of fibrous particles and is used as a fiber source and bulking agent in low-calorie formulations.

Cheese   The product obtained by the coagulation of the milk protein casein by suitable enzymes and/or acid-producing bacteria. The resulting curd, after separation from the whey, is used at once in unripened cheese or is ripened by the action of beneficial bacteria, molds, yeast, and enzymes for ripened cheese.

Cheese Culture   Bacteria used in the coagulation of the milk protein casein in the formation of cheese. It converts milk into cheese curd by the reduction of pH followed by processing to precipitate the protein as a curd.

Cheese Powder   A dry form of cheese prepared by slurrying cheese in water to 35 to 45 percent solids and further processing into a powder form. Cheese powders are water soluble. It is used in instant soups, dry spaghetti sauce, dry sauces, and snack foods.

Chervil   A spice derived from the plant *Anthriscus cerefolium* which is related to the parsley family. It is used in soufflés, sauces, meats, and fish.

Chewing Gum Base   A base, containing masticatory substances such as chicle, used in the manufacture of chewing gum.

Chia Oil   The oil from the seed of any one of the plants belonging to the genus *Salvia*, for example, the shrub *Hispanica* which grows in Mexico. It consists mainly of linolenic and linoleic acid. It is used in the preparation of soft drinks.

Chicle   A natural masticatory substance of vegetable origin which is used in chewing gum base. It is the latex of the sapodilla tree, obtained by cutting the bark to yield the latex which is boiled to remove about two-thirds of the water. The resulting semisolid mass is molded into chicle blocks which form the base for chewing gum.

Chilte   A substance of vegetable origin used as a masticatory substance in chewing gum base.

Chiquibul A substance of vegetable origin used as a masticatory substance in chewing gum base.

Chives A spice from the *Allium schoenoprasum* plant whose slender rush-like green leaves are chopped and used to provide a subtle onion flavor and to enhance food appearance. It is used as a garnish and topping.

Chlorine Cl. A gas used to age and bleach flour.

Chlorine Dioxide ClO_2. A gas used in bleaching and aging flour. It acts on the flour almost instantly, resulting in improved color and dough properties. Because usage levels are low the bleaching action is limited.

Chloropentafluoroethane A propellant and aerating agent for foamed or sprayed foods.

Chlorophyll A colorant that is a green pigment present in all green plants. It is used in sausage casings, oleomargarine, and shortening.

Chocolate A solid or semiplastic food made from chocolate liquor derived from cocoa nibs, which are obtained from the cocoa bean. Chocolate contains more fat and less protein than cocoa. The products derived from chocolate include bitter or plain chocolate; sweet chocolate containing sugar, milk, flavoring, and cocoa butter; and milk chocolate, which is made from sweet or bitter chocolate plus a milk source with or without cocoa butter and flavoring. It is used as a flavor in candy, dairy products, and baked goods.

Chocolate Liquor See **Cocoa Liquor**

Cholic Acid $C_{24}H_{40}O_5$. An emulsifier that exists as colorless plates or a white crystalline powder which has a bitter taste with a sweetish aftertaste. It is slightly soluble in water. It functions as an emulsifying agent in egg white.

Choline A substance grouped as a member of the vitamin B complex, although not a vitamin by definition. It is water soluble and is important in nerve function and fat metabolism. It occurs in egg yolk, beef liver, and grains.

Cider Vinegar The product made by the alcoholic and subsequent acetous fermentation of apple juice or concentrate thereof. It contains not less than 4 g acetic acid in 100 cm³ at 20°C. It has a light to medium amber color. It is used in salad dressings, mayonnaise, and sauces. The term *vinegar* refers to cider vinegar, also termed apple vinegar.

Cinnamon A spice made from the dried bark of the evergreen tree *Cinnamomum cassia*. Commercial types are Saigon Cassia and Batavia Cassia. Ceylon cinnamon is the dried inner bark of shoots of *C. zeylanicum* Nees. In the ground form it is used in beverages, desserts, and fruits while in the stick form it is used in beverages, meats, and fruits.

Cinnamyl Anthranilate $C_{16}H_{15}NO_2$. A synthetic flavoring agent that is a stable, light yellow powder of fruity cherry and grape odor. It should be stored in glass or polyethylene containers. It is used in flavors such as grape, raspberry, and cherry with applications in beverages, candy, and baked goods at 4 to 7 parts per million.

Cinnamyl Isobutyrate A synthetic flavoring agent that is a moderately stable, colorless to light yellow liquid of dry fruity color. It is stored in glass or tin. It is used to give a lift to jasmine with applications in baked goods and candy at 8 parts per million.

Citral $C_{10}H_{16}O$. A liquid flavoring agent, light yellow in color with a citrus odor. It occurs in lemon and lemongrass oils. It is usually obtained from citral-containing oils by chemical means but may also be prepared synthetically. It is soluble in fixed oils, mineral oil, and propylene glycol. It is moderately stable and should be stored in glass, tin, or resin-lined containers. It is used in flavors for lemon with applications in candy, baked goods, and ice cream at 20 to 40 parts per million. It is also termed 2,6-dimethyl-octadian-2-6-al-8.

Citric Acid $C_6H_8O_7$. An acidulant and antioxidant produced by mold fermentation of sugar solutions, by extraction from lemon and lime juice, and from pineapple canning residue. It is the predominant acid in oranges, lemons, and limes. It exists in the anhydrous ($C_6H_8O_7$) and monohydrate ($C_6H_8O_7 \cdot H_2O$) form. The anhydrous form is crystallized in hot solutions and the monohydrate form is crystallized from cold (below 36.5°C) solutions. Anhydrous citric acid has a solubility of 146 g and monohydrate citric acid has a solubility of 175 g per 100 ml of distilled water at 20°C. A 1 percent solution has a pH of 2.3 at 25°C. It is a hygroscopic, strong acid of tart flavor. It is used as an acidulant in fruit drinks and carbonated beverages at 0.25 to 0.40 percent, in cheese at 3 to 4 percent, and in jellies. It is used as an antioxidant in instant potatoes, wheat chips, and potato sticks to prevent spoilage by trapping the metal ions. It is used in combination with antioxidants in the processing of fresh frozen fruits to prevent discoloration.

Citronellyl Propionate $C_{13}H_{24}O_2$. A synthetic flavoring agent that is a moderately stable, colorless liquid of light rose-fruity odor. It is practically insoluble in water but is miscible with alcohol. It is stored in glass or tin containers. It has application in baked goods, candy, beverages, and ice cream at 3 to 19 parts per million.

Citrus Oil A flavorant obtained by pressing the oil from the rind of citrus fruits. It is largely composed of terpenes and sesquiterpenes plus the flavor-imparting oxygenated components. It is partly water soluble, not stable, and is used in beverages.

Clarified Butter See **Butter, Clarified**.

Clear Flour The portion of straight flour (all the flour that can be milled from a wheat blend) that remains after the removal of the patent streams. Clear flours from hard wheat are generally high in ash, dark in color, and high in protein. It is used to increase the strength of flour and is used in rye, dark bread, and pastries.

Clove A spice that is the unripened bud from the clove tree *Eugenia caryophyllata* Thumb. It is very pungent and is used in the whole form in fruit punches, relishes, marinades, and sauces. In the ground form, it is used in cakes, cookies, and meat sauces.

Coarse-Ground Wheat See **Crushed Wheat**.

Cochineal A red colorant extracted from the dried bodies of the female insect *Coccus cacti*. The coloring is carminic acid in which the water-soluble extract is cochineal. It precipitates at pH 3, has good stability at pH 4, and excellent stability at pH 5 to 8. It has low tinctorial strength and has excellent stability to heat and light. It is also stable in retorted protein systems where other food dyes are unstable. It is used in foods requiring red coloring.

Cocoa Butter The fat obtained by pressing chocolate liquor, obtained from roasted cocoa nibs, to yield cocoa butter and presscake. It has a melting point of approximately 33°C but is a hard, brittle solid at room temperature. It is used in the manufacture of coatings for candies, the coatings consisting mainly of mixtures of roasted cocoa nibs, sugar, and cocoa butter. It is also used in confections.

Cocoa Liquor   The liquor obtained by the grinding of cocoa nibs from the cocoa bean. The liquor converts into cocoa powder and cocoa butter as end products. It is a primary ingredient in chocolate manufacture. It is also termed chocolate liquor.

Cocoa Powder   The powder produced by the grinding, pulverizing, and air classification of the cocoa presscake which is obtained by compressing the cocoa liquor, obtained from the cocoa nibs, into a presscake and cocoa butter. There are two main types of powder — alkalized and natural. The alkalized (Dutch processed) has a pH range of 6.5 to 8.1; a red-brown shade which tends to develop red-brown end products; a mild flavor; and is used in beverages, retail cocoa powder, puddings, and ice cream. The natural has a pH range of 5.2 to 5.9; a yellow-orange color with a tendency to produce light brown end products; and is used in the baking industry to impart color and flavor; and also used in candy, syrups, and toppings.

Coconut   The nut obtained from the coconut palm. It provides a source of coconut meat and coconut oil.

Coconut, Desiccated   The dried coconut meat whose reduced moisture content increases its stability. It is available in various shapes and sizes. It is used to impart flavor in desserts, baked goods, and candies.

Coconut Oil   The oil obtained from the kernel of the nuts of the coconut palm. It has a sharp melting character (narrow plastic range) in that it changes abruptly from a hard, brittle solid to a clear oil within a temperature change of a few degrees and the transition occurs at room temperature range. It melts at 25°C and is more completely solid than butter at 10°C. These properties make it suited for the preparation of shortenings where brittleness and a large change in consistency with a small temperature change are undesirable. Partially hydrogenated coconut oil has hydrogen added to part of the unsaturated carbon bonds to provide a more solid consistency. It is used in confections, baked goods, and margarine.

Collagen   A protein that is the principal constituent of connective tissue and bones of vertebrates which can be converted to gelatin and glue by boiling in water.

Copper Cu. A metal necessary for the maintenance of normal erythropoiesis and the prevention of iron deficiency anemia, iron being essential in hemoglobin synthesis.

Copper Gluconate $C_{12}H_{22}CuO_{14}$. A light blue powder used as a dietary supplement.

Copra A dried coconut meat used as a flavorant. It provides a stable form of coconut oil and meat to allow for storage. It usually has 63 to 68 percent oil.

Coriander A spice that is the dried, ripe fruit of *Coriandrum sativum* L. It has a pleasing, aromatic taste. It is used in sausage, variety meats, and curry powder in the ground form, and in pickles, baked goods, and stuffing in the whole form.

Corn The maize grain, which is the source of various ingredients. It is used in the kernel form for food; it is dry milled into flour, grits, and meal; and it is wet milled into starches, dextrins, dextrose, and other by-products. The kernel consists of four basic parts which are the starch section, the corn germ, gluten, and hull. The starch section comprises approximately 61 percent of the kernel, while the corn germ comprises approximately 4 percent of the kernel. The term *corn* refers to other cereal crops in different areas of the world.

Corn Bran A dry-milled product of high fiber content obtained from corn. It can be used to increase the fiber content of breads, cookies, cereals, and to thicken gravies and soups.

Corn Flour A finely ground flour made from milling and sifting maize or obtained as a by-product of cornmeal. It is used as pancake flour.

Cornmeal A ground corn of specified mesh profile that is made from white or yellow maize. It is used in corn bread mix.

Corn Oil The oil obtained from the germ of the maize plant. The unsaturated fatty acids linoleic and oleic make up 80 to 85 percent of the total fatty acids. The tocopherols prevent the oil from oxidizing rapidly. It has a low melting point of -18 to $-10°C$. It is used in mayonnaise, margarine, salad oil, and bakery products.

Cornstarch The starch made from the endosperm of corn and containing amylose and amylopectin starch molecules. When starch is heated in water it forms a viscous, opaque paste. The paste forms semisolid gels upon cooling and has the ability to form strong adhesive films when spread and dried. Cornstarch is not freeze-thaw stable and is used widely except when clarity or the lack of gel formation is desired. It exists as fine or coarse powders. The coarse starch is sometimes termed pearl starch. It is used in sauces, gravies, puddings, pie fillings, and salad dressings. The typical usage level is 1 to 5 percent. It is also termed common, regular, or unmodified cornstarch.

Cornstarch, Acid-Modified   A starch produced by treating suspended cornstarch in water with dilute mineral acid at high temperatures for varying time periods. This is followed by neutralization with sodium carbonate upon obtainment of the desired viscosity. This produces starches that have decreased viscosity when warm but still form gels when cooled. It is used in the manufacture of starch-based gum candies. It is also termed thin-boiling starch, and acid-modified cornstarch. Also, esters and ethers can be formed in which only one end of the addition molecule is attached to the starch molecule. These starches have freeze-thaw stability, shear resistance, and acid resistance, and are used in sauces, gravies, and frozen foods.

Cornstarch, Oxidized   See **Oxidized Cornstarch**.

Corn Sugar   See **Dextrose**.

Corn Sugar Vinegar   The product made by the alcoholic and subsequent acetous fermentation of corn sugars according to federal regulations. It is of amber color and has a minimum of 4 percent acid (expressed as acetic acid). It functions as an acidulant in foods.

Corn Syrup   A corn sweetener that is a viscous liquid containing maltose, dextrin, dextrose, and other polysaccharides. It is obtained from the incomplete hydrolysis of cornstarch. It is classified according to the degree of conversion which is expressed as the dextrose equivalent (DE), which is the measure of sweetness of the corn syrup as compared to that of a sucrose syrup. Generally the greater the degree of conversion, the sweeter the syrup. Corn syrup is used as a replacement for sucrose but is less sweet than sucrose. It can control crystallization in candy making, contribute body in ice cream, and provide pliability in confections. It is also termed Glucose Syrup.

Corn Syrup Solids   The dry form of corn syrup used where it is impractical to use the liquid syrup. See **Corn Syrup**.

Cottonseed Oil   The oil obtained from the seeds of the cotton plant. It is a semidrying oil which resists the development of rancidity. It requires removal of stearine to prevent clouding at refrigerated temperatures. It is used in shortening, cooking oil, salad oil, and margarine.

Coumarin $C_9H_6O_2$. A flavorant with an odor similar to vanilla. It is found as a component in many imitation vanillas. It can be used by itself, but it tends to dominate in vanilla.

Cracked Wheat   The wheat prepared by cracking or cutting cleaned wheat, other than durum wheat and red durum wheat, into angular fragments. The proportions of the natural constituents, other than moisture, remain unaltered. The moisture content does not exceed 15 percent.

Cracker Flour   Flour that is long patented or straight grades of soft wheat flour, containing 9 to 10.5 percent protein. It is of low absorption and has short mixing requirements. See **Flour category**.

Cranberry Extract   A natural red colorant with good pH stability, and fair heat, light, and chemical stability. The anthocyanidin pigments in cranberry are peonidin and cyanidin. The extract has low tinctorial strength and good stability at pH 3 to 4. Because the color is affected by pH, it can only be used in acidic mediums such as beverages.

Cream   That portion of milk that is high in milkfat and will rise to the top of undisturbed milk. It is obtained by the separation of the fat fraction of the milk to concentrations ranging from 18 to 40 percent fat. Cream is labeled according to the fat content: heavy whipping cream has a minimum of 36 percent fat; light whipping cream has 30 to 36 percent fat; and light, coffee, or table cream has 18 to 30 percent fat. The lower fat creams are usually prepared by blending a high-fat cream with milk. Cream is used in ice cream mix, whipped toppings, and sauces.

Cream of Tartar $KHC_4H_4O_6$. The acid potassium salt of tartaric acid occurring as crystals or powder. It is relatively poorly soluble having a solubility in 100 ml of water of 0.8g at 25°C and 6.1 g at 100°C. A 1 percent solution at 30°C has a pH of 3.4. Chemical names are potassium acid tartrate, potassium hydrogen tartrate, and potassium bitartrate. It functions to complex with heavy metal ions and to regulate pH; it can have a gentle laxative action at given adequate levels. The acidulant is used in chemical leavening to release carbon dioxide which produces the loaf volume. It has limited reactivity in the cold so when used in reduced-temperature batters it has little gas evolution during the initial mixing. At room temperatures, it has a relatively fast reaction rate. It functions as a taste regulator in sugar icing and in the controlled crystallization of toffees and fondants by the regulated inversion of sucrose. It is used in baked goods, crackers, candy, and puddings.

Crown Gum   A product of vegetable origin used as a masticatory substance in chewing gum base.

Crushed Wheat   The wheat prepared by crushing cleaned wheat, other than durum wheat and red durum wheat, with the proportions of the natural constituents, other than moisture, remaining unaltered. The moisture content does not exceed 15 percent. It is also termed coarse-ground wheat.

Cumin   A spice that is the dried, ripe fruit of *Cuminum cyminum* L. It is usually obtained in the ground form. It has a warm, pleasant, balsamic flavor. It is used in cheese, soups, relishes, and meats.

Cupric Aspartate   A source of copper.

Cupric Carbonate   A source of copper.

Cupric Chloride   A source of copper.

Cupric Citrate   A source of copper.

Cupric Glycerophosphate   A source of copper.

Cupric Nitrate   A source of copper.

Cupric Nucleinate   A source of copper.

Cupric Pyrophosphate   A source of copper.

Cupric Sulfate   A source of copper.

Cuprous Aspartate   A source of copper.

Cuprous Carbonate   A source of copper.

Cuprous Chloride A source of copper.

Cuprous Citrate A source of copper.

Cuprous Glutamate A source of copper.

Cuprous Glycerophosphate A source of copper.

Cuprous Iodide A source of copper and iodine used in table salt.

Cuprous Nitrate A source of copper.

Cuprous Nucleinate A source of copper.

Cuprous Pyrophosphate A source of copper.

Cuprous Sulfate A source of copper.

Curry Powder A blend of spices used as seasoning in curries, sauces, and meats. Typical spices in the blend include coriander, ginger, nutmeg, clove, cinnamon, red pepper, and onion salt.

Cyanocobalamin $C_{63}H_{88}CoN_{14}O_{14}P$. Vitamin B_{12}, a water-soluble vitamin required for the normal development of red blood cells. Its deficiency causes pernicious anemia. It is stable in neutral conditions and is more stable for storage than for processing conditions. It is found in meat, fish, and milk.

Cyclamic Acid A flavor potentiator that is soluble in water and has a sweet taste. It is about 30 times as sweet as sugar.

Cyclohexyl Acetate A synthetic flavoring agent that is a stable, colorless liquid of fruity odor. It is stored in glass, tin, or resin-lined containers. It is used for flavors such as apple, banana, blackberry, and raspberry with applications in beverages, ice cream, candy, and baked goods at 15 to 110 parts per million.

Cyclohexyl Butyrate A synthetic flavoring agent that is a stable, colorless liquid of fruity odor. It should be stored in glass, tin, or resin-lined containers. It is used in pineapple, apricot, banana, and berry flavor with applications in beverages, ice cream, and candy at 4 to 9 parts per million.

Cyclohexyl Cinnamate   A synthetic flavoring agent that is a stable, colorless to light yellow liquid of fruity odor. It is stored in glass or tin containers. It is used in peach and cherry flavors with applications in ice cream, candy, and baked goods at 5 to 20 parts per million.

Cyclohexyl Formate   A synthetic flavoring agent that is a moderately stable, colorless liquid of fruity odor. It should be stored in glass or tin containers. It is used in apple and plum flavors and gives a lift to fruity flavors. It has application in beverages, candy, ice cream, and baked goods at 3 to 11 parts per million.

Cyclohexyl Propionate   A synthetic flavoring agent that is a stable, colorless liquid of fruity odor. It is stored in glass or tin containers. It is used in fruit flavors such as pineapple with applications in beverages, candy, ice cream, and baked goods at approximately 3 parts per million.

Cydonia Seed   See **Quince Seed**.

Cysteine $C_3H_7NO_2S$. A nonessential amino acid that functions as a nutrient and dietary supplement. It is used in foods to prevent oxygen from destroying vitamin C and is used in doughs to reduce mixing time.

Cystine $C_6H_{12}N_2O_4S_2$. A nonessential amino acid that acts as a nutrient and dietary supplement. It is very slightly soluble in water and in alcohol. It improves the biological quality of the total protein in foods containing naturally occurring intact protein.

Danish Agar See **Furcellaran**.

Decanol A flavoring agent that is a stable, colorless liquid which has a floral odor resembling orange flowers. It is soluble in alcohol and insoluble in water. It has application in beverages, candy, baked goods, and ice cream at approximately 2 to 5 parts per million. It is also termed decyclic alcohol.

Decyclic Alcohol See **Decanol**.

Dehydroacetic Acid (DHA) $C_8H_8O_4$. A preservative that is a crystalline powder with a solubility of less than 0.1 g in 100 g H_2O at 25°C. It can undergo a variety of chemical reactions which give it utility in many applications. It is used at 0.01 to 0.5 percent for microbiological growth inhibition in various foods. It is used for cut or peeled squash, with no more than 65 parts per million remaining in or on the prepared squash.

Deoxycholic Acid $C_{24}H_{40}O_4$. An emulsifier that is a white crystalline powder, being practically insoluble in water. It is used in dried egg whites.

Dextran $(C_6H_{10}O_5)_x$. A gum obtained by fermentation of sugar with bacteria of the genus *Leuconostoc*. It has little commercial value and is used mainly as a blood plasma extender. Other possible applications are baked goods, confectionary, and as a preservative coating for foods.

Dextrin   A carrier formed from the treatment of starch by dry heat, acid, or enzymes. It can be formed from amylose and amylopectin-type starches. The cold-water soluble dextrins are used as carriers for active ingredients such as flavors. Industrially, it is used as an adhesive. It is used as a flavor carrier in dry-mix beverages, soups, and gravy.

Dextrose $C_6H_{12}O_6$. A corn sweetener that is commercially made from starch by the action of heat and acids or enzymes, resulting in the complete hydrolysis of the cornstarch. There are two types of refined dextrose commercially available. Dextrose hydrate, containing 9 percent by weight of water of crystallization and being the most often used, and anhydrous dextrose, which contains less than 0.5 percent water. Corn sugar is a reducing sugar and produces a high-temperature browning effect in baked goods. It is used in ice cream, bakery products, and confections. It is also termed glucose and corn sugar.

Diacetyl Tartaric Acid Esters of Mono- and Diglycerides   A hydrophilic emulsifier used in oil-in-water emulsions. The connecting of glycerol with tartaric acid prior to esterification of the other part of the glycerol increases the hydrophilicity of the emulsifier. It functions as a dough conditioner in freestanding breads and rolls to strengthen the gluten which improves crumb softness, crust, and increased volume. It is used in coffee whiteners for dispersion. It is used in chocolate couverture to adjust the consistency, viscosity, and adhesion ability. In reduced-calorie breads, it reduces the quantity of shortening required and maintains volume. It is also termed acetylated tartaric acid monoglyceride, and acetyl tartrate mono- and diglyceride.

Diammonium Phosphate $(NH_4)_2HPO_4$. A leavening agent used in the production of cookies and crackers. It is used in yeast foods to supply nitrogen and phosphorus; in crackers it functions as a buffer to prevent excess alkalinity upon addition of soda.

Dibasic Calcium Phosphate, Anhydrous   See **Dicalcium Phosphate, Anhydrous**.

Dibasic Calcium Phosphate, Dihydrate   See **Dicalcium Phosphate, Dihydrate**.

Dicalcium Phosphate, Anhydrous $CaHPO_4$. A mineral supplement
and dough conditioner. It contains approximately 29 percent calcium.
It is practically insoluble in water, with a solubility of 0.02 g per 100
ml H_2O at 25°C. It is also termed calcium phosphate, dibasic anhydrous. It is used as a mineral supplement in prepared breakfast cereals, enriched flour, and noodle products.

Dicalcium Phosphate, Dihydrate $CaHPO_4 \cdot 2H_2O$. A source of calcium and phosphorus that functions as a dough conditioner and
bleaching agent. It functions as a dough conditioner in bakery products, as a bleaching agent in flour, as a source of calcium and phosphorus in cereal products, and as a source of calcium for alginate gels. It
contains approximately 23 percent calcium. It is practically insoluble
in water. It is also termed dibasic calcium phosphate, dihydrate; and
calcium phosphate dibasic, hydrous. It is used in dessert gels, baked
goods, cereals, and breakfast cereals.

Diglyceride   A lipophilic emulsifier prepared by direct esterification
of two fatty acids with glycerol, or by interesterification between
glycerol and other triglycerides. It often occurs as a blend with
monoglycerides. It is widely used in numerous foods such as ice
cream, puddings, margarine, doughs, shortenings, peanut butter, and
coffee whiteners. It has numerous functions including the provision of
dough conditioning, prevention of fat separation, and the provision of
emulsion stability and dispersibility.

Dilauryl Thiodipropionate (DLTDP) $C_{30}H_{58}O_4S$. An antioxidant that
exists as white crystalline flakes of sweetish ester-like odor. It is
insoluble in water but soluble in inorganic solvents. It is used in fats
and oils to prevent rancidity. It is used in peanut oil at a maximum
usage level of 200 mg/kg.

Dill Seed   A spice that is the dried, ripe fruit of the plant *Anethum
graveolens* L. It is extremely pungent and slightly dominant. It is used
in dips, spreads, sauces, and meats.

Dill Weed   A spice made from the leaf of the dill plant. While dill seed
has a camphorous, slightly bitter taste and fragrance, the weed has a
delicate bouquet which enhances rather than dominates. It is used in
meats and sauces.

2,6-Dimethyl-Octadian-2-6-al-8   See **Citral**.

Dimethylpolysiloxane   An antifoaming agent used in fats and oils. It prevents foaming and spattering when oils are heated and prevents foam formation during the manufacture of wine, refined sugar, gelatin, and chewing gum. It is also termed methylpolysilicone and methyl silicone.

Dioctyl Sodium Sulfosuccinate $C_{20}H_{37}NaO_7S$. A wetting and emulsifying agent that is slowly soluble in water, having a solubility of 1 g in 70 ml of water. It functions as a wetting agent in fumaric acid-containing powdered fruit drinks to help the acid dissolve in water. It is used as a stabilizing agent on gums at not more than 0.5 percent by weight of the gum. It is used as a flavor potentiator in canned milk where it improves and maintains the flavor of the sterilized milk during storage. It also functions as a processing aid in the manufacture of unrefined sugar.

Dipotassium Monohydrogen Orthophosphate   See **Dipotassium Phosphate**.

Dipotassium Monophosphate   See **Dipotassium Phosphate**.

Dipotassium Phosphate K_2HPO_4. The dipotassium salt of phosphoric acid which functions as a stabilizing salt, buffer, and sequestrant. It is mildly alkaline with a pH of 9 and is soluble in water with a solubility of 170 g per 100 ml of water at 25°C. It improves the colloidal solubility of proteins. It acts as a buffer against variation in pH and, for example, it is used as a buffer in coffee whiteners against pH variation in hot coffee and to prevent feathering. It functions as an emulsifier in specified cheeses and as a buffering agent for processed foods. It is also termed dipotassium monohydrogen orthophosphate, potassium phosphate dibasic, and dipotassium monophosphate.

Disodium Calcium EDTA   A sequestrant and chelating agent whose complete name is disodium calcium ethylenediamine tetraacetate. It is a nonhygroscopic powder that is colorless, odorless, and tasteless at recommended use levels. A 1 percent solution has a pH of 6.5 to 7.5. It is used to control the reaction of trace metals with some organic and inorganic components in food; to prevent deterioration of color, texture, and development of precipitates; and to prevent oxidation. Its function is comparable to that of disodium dihydrogen EDTA. See **EDTA**.

Disodium Dihydrogen EDTA   A sequestrant and chelating agent whose complete name is disodium dihydrogen ethylenediamine tetraacetate. It is a nonhygroscopic powder that is colorless, odorless, and tasteless at recommended use levels. A 1 percent solution has a pH of 4.3 to 4.7. It is used to control the reaction of trace metals; to include calcium and magnesium with other organic and inorganic components in food to prevent deterioration of color, texture, and development of precipitates; and to prevent oxidation. Its function is comparable to that of calcium disodium EDTA. See **EDTA**.

Disodium Dihydrogen Pyrophosphate   See **Sodium Acid Pyrophosphate**.

Disodium Diphosphate   See **Sodium Acid Pyrophosphate**.

Disodium 5′-Guanylate (DSG) $C_{10}H_{12}N_5Na_2O_8P \cdot xH_2O$. A flavor enhancer that belongs to the same family as monosodium glutamate and disodium inosinate and is often used together with them. It is 50 to 100 times more potent than monosodium glutamate which is primarily used as a meaty flavor enhancer. Disodium guanylate enhances a variety of flavors, modifies salty or sweet flavors, and suppresses undesirable flavors. It is soluble in water and creates a sense of increased viscosity in liquid foods. It is reasonably stable to changes in pH but is not stable in the presence of phosphatase enzymes. It is made from yeast ribonucleic acid. It is used in soups, sauces, snack items, and canned foods at a typical usage level of 0.01 to 0.05 percent.

Disodium 5′-Inosinate $C_{10}H_{11}N_4Na_2O_8P \cdot xH_2O$. A flavor enhancer which performs as does disodium guanylate, but only when present at approximately twice the level. See **Disodium Guanylate**.

Disodium Malate   An acidulant.

Disodium Monohydrogen Orthophosphate   See **Disodium Phosphate**.

Disodium Monohydrogen Orthophosphate Dihydrate   See **Disodium Phosphate**.

Disodium Monophosphate   See **Disodium Phosphate**.

Disodium Phosphate Na_2HPO_4. The disodium salt of phosphoric acid which functions as a protein stabilizer, buffer, dispersant, and coagulation accelerator. It is mildly alkaline with a 1 percent solution having a pH of 9.2. It is moderately soluble in water with a solubility of 12 g in 100 ml at 25°C. It is used in farina and macaroni to shorten the cooking time by making the particles swell faster and cook more thoroughly. In evaporated milk it acts as a buffer and prevents gelation, also acting as a buffer in coffee whiteners. It is an accelerator of the setting time in instant puddings. In cream sauce and whipped products it functions as a dispersant by producing a swelling of protein. It is also termed disodium monohydrogen orthophosphate; sodium phosphate, dibasic; and disodium monophosphate.

Disodium Phosphate, Dihydrate   See **Disodium Phosphate, Duohydrate**.

Disodium Phosphate, Duohydrate $Na_2HPO_4 \cdot 2H_2O$. An emulsifier, buffer, and mineral supplement. It is moderately soluble in water with a solubility of 15 g in 100 ml of water at 25°C. A 1 percent solution has a pH of 9.1. It is used in processed cheese for uniform texture and smoothness. It is also termed disodium phosphate, dihydrate and sodium phosphate dibasic, dihydrate.

Disodium Tartrate   See **Sodium Tartrate**.

Distilled Monoglyceride   An emulsifier containing a minimum of 90 percent monoglyceride derived from edible fat and glycerine. It is an active monoglyceride produced by distillation to obtain the monoglyceride fraction, which is the part that functions as an emulsifier or food quality improver. Commercially termed monoglycerides also contain diglycerides, triglycerides, and so on. It is used in margarine, peanut butter, shortenings, bakery goods, and whipped desserts to improve texture and consistency. Typical usage levels are 0.1 to 1.0 percent.

Distilled Vinegar   See **Vinegar, Distilled**.

2,6-Di-Tert-Butyl-Para-Cresol   See **Butylated Hydroxytoluene**.

Dodecyl Gallate   An antioxidant used in cream cheese, instant mashed potatoes, margarine, fats, and oils.

Dolomitic Limestone   See **Limestone, Dolomitic**.

Dough Conditioner   A blend of minerals used in baked goods. It is usually contained within yeast foods as a blend of calcium salts, sulfates, and phosphates which toughen the gluten. Usage of hard water generally results in better breads so the minerals serve to minimize the effect of variables in water conditions. It is also termed yeast food.

Dried Buttermilk See **Buttermilk, Dried**.

Dried Milk See **Whole Milk Solids**.

Dried Skim Milk See **Milk Solids – Not-Fat**.

Dry Ice See **Carbon Dioxide**.

Dry Whole Milk See **Milk Powder**.

Durum Flour The fine powder obtained from durum wheat, which is fine enough to pass through a number 100 U.S. sieve. It is used principally in macaroni and spaghetti products because it provides the desired texture and consistency. See **Durum Wheat**.

Durum Granular The product obtained from durum wheat by grinding to obtain semolina to which flour is added so that 7 to 20 percent passes through a number 100 U.S. sieve. It is used in macaroni and spaghetti. See **Durum Wheat**.

Durum Semolina See **Semolina**.

Durum Wheat The wheat obtained from the durum wheat kernel. It differs from other hard wheats in that the starch swelling capacity is greater and the gluten has different characteristics which result in tough, elastic doughs. As compared to hard wheat dough, it can be extruded through a small hole at lower pressure and in breads results in lower loaf volume. It is used almost exclusively in macaroni and spaghetti products because it is easily processed to produce a smooth, mechanically strong product of desired color which when cooked will maintain its shape and be of firm consistency. Products derived from the wheat include durum flour, durum granular, and durum semolina.

E

EDTA The abbreviation for ethylenediaminetetraacetate, a sequestrant and chelating agent that functions in water but not fats and oils. It is used to control the reaction of trace metals with some organic and inorganic components to prevent deterioration of color, texture, and development of precipitates, as well as to prevent oxidation which results in rancidity. The reactive sites of the metal ions are blocked, which prevents their normal reactions. The most common interfering metal ions in food products are iron and copper. It can be used in combination with the antioxidants BHT and propyl gallate. It is used in margarine, mayonnaise, and spreads to prevent the vegetable oil from going rancid. It is used in canned corn prior to retorting in preventing discoloration caused by trace quantities of copper, iron, and chromium. It also inhibits copper-catalyzed oxidation of ascorbic acid. It occurs as calcium disodium EDTA, and disodium dihydrogen EDTA. Its use is approved in specified foods, with an average usage level being in the range of 100 to 300 parts per million.

Egg The hard-shelled reproductive body of poultry protected by a shell, largely composed of calcium carbonate, which represents approximately 11 percent of its total weight. Inside the shell are the shell membranes, which are principally protein. The yolk, which represents approximately 31 percent of the egg's weight, contains protein, fat, and all the known vitamins except vitamin C. Most of the egg's calories come from the yolk. The egg white is protein and represents approximately 58 percent of the weight. The white does not appear white in color until beaten or cooked. There is a thick and thin white, which consists mainly of ovalbumin, conalbumin, ovoglobulin, ovomucoid, and ovomucin. Eggs are used whole, as egg white, as yolk, or any combination thereof. They are used for coagulation, foam formation, emulsification, nutrition, flavor, and color.

Egg Albumen   The protein fraction of egg, which is also termed egg white. It represents approximately 65 percent of the edible egg and is composed of approximately 87 percent water, 11 percent protein, and 1 percent carbohydrate. It provides a source of protein and provides foam upon whipping. It is used in meringues, cakes, and desserts.

Egg White   See **Egg Albumen**.

Egg Yolk   The yellow portion of the egg, representing approximately 35 percent of the edible egg. It is composed of approximately 49 percent water, 16 percent protein, 32 percent fat, and trace carbohydrate. It is used as an emulsifier in mayonnaise, salad dressing, and cream puffs. It is also used as a source of color.

Enriched Bleached Flour   Flour that has been whitened by removal of the yellow pigments and fortified with vitamins and minerals. The added vitamins are thiamin, riboflavin, niacin, or niacinamide, and may include vitamin D. The minerals are iron and may include calcium. It is used in baked goods.

Entire Wheat Flour   See **Whole Wheat Flour**.

Epsom Salt   See **Magnesium Sulfate**.

Ergosterol   A steroid alcohol that when irradiated with ultraviolet light yields calciferol (Vitamin D_2). Irradiated ergosterol is added to milk for vitamin D fortification.

Erythorbic Acid $C_6H_8O_6$. A food preservative that is a strong reducing agent (oxygen accepting) which functions similarly to antioxidants. In the dry crystalline state it is nonreactive, but in water solutions it reacts readily with atmospheric oxygen and other oxidizing agents, making it valuable as an antioxidant. During preparation, dissolving and mixing should incorporate a minimum amount of air and storage should be at cool temperatures. It has a solubility of 43 g per 100 ml of water at 25°C. One part erythorbic acid is equivalent to 1 part ascorbic acid and equivalent to 1 part sodium erythorbate. It is used to control oxidative color and flavor deterioration in fruits at 150 to 200 parts per million. It is used in meat curing to speed and control the nitrite curing reaction and prolong the color of cured meat at levels of 0.05 percent.

Erythroascorbic Acid   An antioxidant used in combination with citric acid to prevent the browning of bananas during freezing and defrosting.

Erythrosine   See **FD&C Red #3**.

Ester Gum   A density adjuster prepared from glycerol of non-animal sources and refined wood rosin of pine trees. It is used to adjust the specific gravity of the citrus oil or oil phase to be similar to the specific gravity of the beverage emulsion and thus prevent the oil from rising or settling in the finished beverage. It also imparts some cloudiness. It is soluble in aromatic and petroleum hydrocarbons, terpenes, esters, ketones, citrus, and essential oils. It is used in lemon and orange drinks and also as a masticatory substance in chewing gum base. It is technically termed glyceryl abietate and is also called glycerol ester of wood rosin and glycerol dihydroabietate.

Ethoxyquin $C_{14}H_{19}NO$. An antioxidant used in the preservation of color in chili powder, ground chili, and paprika.

Ethoxylated Mono- and Diglycerides   An emulsifier prepared by the glycerolysis of edible vegetable fats and reacting with ethylene oxide. It is hydrophilic, being soluble in water and partially soluble in oil. It contributes to freeze thaw stability and overrun in whipped toppings. It functions as a dough conditioner/emulsifier in baked goods, as an emulsifier in coffee whiteners, icings, and frozen desserts. Typical usage levels are 0.20 to 0.45 percent. It is also termed polyglycerate 60 and polyoxyethylene (20) mono- and diglycerides of fatty acids.

Ethyl Crotonate   A synthetic flavoring agent that is a moderately stable, colorless to light yellow liquid of sharp winey note. It should be stored in glass, tin, or resin-lined containers. It is used in fruit flavors for application in baked goods, beverages, and candy at 2 to 7 parts per million.

Ethyl-2,4-Hexadienoate   See **Ethyl Sorbate**.

Ethyl Isobutyrate   A synthetic flavoring agent that is a stable, colorless liquid of dry, fruity odor. It should be stored in tin, glass, or resin-lined drums. It is used to give fruity effects to flavors for applications in candy, baked goods, and beverages at 10 to 100 parts per million.

Ethyl Lactate   A solvent manufactured from L (+) lactic acid which is miscible in water and most organic solvents, and is cleared for use as a flavoring agent. It is a naturally occurring constituent of California and Spanish sherries. It is a clear, colorless, nontoxic liquid of low volatility, having a pH of 7 to 7.5. It is used as a food and beverage flavoring agent.

Ethyl Maltol   A synthetic flavor enhancer related to maltol. As compared to maltol, it is two to six times as effective, more soluble (1 g in 55 ml of water at 25°C), and more volatile, with a melting range of about 90°C. It does not contribute a flavor of its own, but modifies the inherent flavor. It is used in beverages to enhance the fruitiness. It is used to increase the richness and smoothness of chocolate flavor and for richer flavor in desserts. The typical usage level range is 1 to 100 parts per million.

Ethyl Methyl Phenyl Glycidate $C_{12}H_{14}O_3$. A synthetic flavoring agent that is a glycidic acid ester. It is a colorless to pale yellow liquid of a strong fruit odor suggestive of strawberries. It is unstable to alkali and moderately stable to weak organic acids. It should be stored in glass, tin, or aluminum containers. It is soluble in fixed oils and in propylene glycol. It is used in flavors for strawberry note and has application in candy, beverages, and ice cream at 6 to 20 parts per million. It is also termed aldehyde C-16.

Ethyl Nonanoate $C_{12}H_{22}O_2$. A synthetic flavoring agent that is a stable, colorless liquid of fruit cognac odor. It is practically insoluble in water and is miscible with alcohol and propylene glycol. It should be stored in glass or tin containers. It is used in flavors such as apple, pear, and cognac with applications in beverages, ice cream, candy, and alcohol beverages at 4 to 20 parts per million.

Ethyl Oxyhydrate   See **Rum Ether**.

Ethyl Paraben   See **Parabens**.

Ethyl Sorbate   A synthetic flavoring agent that is a moderately stable, light yellow liquid of fruity odor. It should be stored in glass or tin containers. It is used in flavors such as pineapple, papaya, and passion fruit with applications in ice cream, beverages, candy, and baked goods at 6 to 18 parts per million. It is also termed ethyl-2, 4-hexadienoate.

Ethyl Vanillin $C_9H_{10}O_3$. A flavoring agent that is a synthetic vanilla flavor with approximately three and one-half times the flavoring power of vanillin. It has a solubility of 1 g in 100 ml of water at 50°C. It is used in ice cream, beverages, and baked goods.

Ethylenediaminetetraacetate See **EDTA**.

Eugenol $C_{10}H_{12}O_2$. A flavoring obtained from clove oil and also found in carnation and cinnamon leaves. It is a stable, light yellow-green liquid of clove odor. It is slightly soluble in water and miscible in alcohol. It should be stored in glass or tin, avoiding iron containers. It is used in spice oils for application in condiments and meats at 100 to 200 parts per million and in baked goods and candy at approximately 30 parts per million.

Extract An alcohol or alcohol-water solution that contains a flavoring ingredient obtained from a spice or some other ingredient which is used as a flavorant. It is used in baked goods, beverages, and ice cream.

Extract of Malted Barley and Corn See **Malted Cereal Syrup**.

Family Flour   See **All-Purpose Flour**.

Farina   Wheat, other than durum or red durum wheat, from which the bran and most of the germ has been removed. It is ground so that not more than 3 percent passes through a number 100 U.S. sieve.

Fast Green FCF   See **FD&C Green #3**.

Fat   Water-insoluble material of plant or animal origin, consisting predominantly of glyceryl esters of fatty acids (triglycerides). Fat ordinarily refers to triglycerides that are semisolid at room temperature. Fat in its liquid state is called oil.

Fatty Acids   Aliphatic acids that may be saturated or unsaturated, consisting of a mixture of certain monobasic carboxylic acids and their associated fatty acids. Fatty acids plus glycerol result in a fat characterized by the fatty acid components. A fatty acid may be used as a lubricant, a binder, a food processing defoamer, and an emulsifier.

FD&C Blue #1   A colorant. It has a solubility in water of 20 g in 100 ml at 25°C. It has good pH stability with only slight fading after one week at pH 3, 7, 8 but is unstable in alkalis such as 10 percent sodium carbonate and 10 percent ammonium hydroxide. It has good stability in 10 percent sugar systems. It has fair stability to light, fair to poor stability to oxidation, and good stability to heat. It has a greenish-blue hue with excellent tinctorial strength. It is used with other primary colors to produce a variety of shades, for example, in combination with FD&C Yellow #5, it gives green. It has good compatibility with food components. It is used in candies, baked goods, soft drinks, and desserts. The common name is brilliant blue FCF.

FD&C Blue #2 A colorant. It has poor pH stability in that after one week at pH 3 and 5 it will appreciably fade, at pH 7 considerably fade, and at pH 8 it fades completely. It is the least soluble of all food colors, with a solubility of 1.6 g in 100 ml of water at 25°C. Complete fading occurs in alkalis such as 10 percent sodium carbonate and 10 percent sodium hydroxide, with fading also occurring in 10 percent sugar systems. It has very poor light stability and oxidation stability, and moderate stability to heat. It has a deep blue hue with poor tinctorial strength. It is the only food color that has good resistance to reducing agents, but has very poor compatability with food components. The major use is in pet food, but it is also used in candies, confections, and baked goods. The common name is indigotine.

FD&C Green #3 A colorant. It has good pH stability, showing after one week a slight fade at pH 3, a very slight fade at pH 5 to 7, and slight fade and appreciably bluer color at pH 8. It has excellent solubility in water with a solubility of 20 g in 100 ml at 25°C. It has fair to good stability to light, poor stability to oxidation, and shows no appreciable change in 10 percent sugar systems. It has a bluish-green hue, with excellent tinctorial strength. It has good compatibility with food components and is occasionally used in cereals, soft drinks, beverages, and desserts. The common name is fast green FCF.

FD&C Red #3 A colorant. It is not recommended for use below pH 5.0, being insoluble at pH 3 to 5 but being stable at pH 7 and 8. It has a solubility in water of 9 g per 100 ml at 25°C. It has fair stability to oxidation and poor to fair stability to light, while having good stability in 10 percent sugar systems. It has exceptional clarity and brilliance, having a bluish-pink hue with very good tinctorial strength. It has poor compatibility with food components and is used in candies and confections as well as cherry dyeing. The common name is erythrosine.

FD&C Red #4 A colorant. It has a solubility in water of 11 g per 100 ml at 25°C. It has fair stability to oxidation, very good stability to light, and good stability to pH changes. It has a yellow-red hue and very good tinctorial strength. It is approved only for use in maraschino cherries. The common name is ponceau SX.

FD&C Red #40   A colorant. It has good stability to pH changes from pH 3 to 8, showing no appreciable change. It has excellent solubility in water with a solubility of 22 g per 100 ml at 25°C. It has very good stability to light, fair to poor stability to oxidation, good stability to heat, and shows no appreciable change in stability in 10 percent sugar systems. It has a yellowish-red hue and has a very good tinctorial strength. It has very good compatibility with food components and is used in beverages, desserts, candy, confections, cereals, and ice cream. The common name is Allura® red AC.

FD&C Yellow #5   A colorant. It has good stability to changes in pH, showing no appreciable change at pH 3 to 8. It has excellent solubility in water with a solubility of 20 g in 100 ml at 25°C. It has good stability to light and heat, fair stability to oxidation, and shows no appreciable change in 10 percent sugar systems. It has a lemon-yellow hue and has good tinctorial strength. It has moderate compatibility with food components and is used in beverages, baked goods, pet foods, desserts, candy, confections, cereal, and ice cream. The common name is tartrazine.

FD&C Yellow #6   A colorant. It has good stability to changes in pH, showing no appreciable change at pH 3 to 8. It has excellent solubility in water with a solubility of 19 g in 100 ml at 25°C. It has moderate stability to light, fair stability to oxidation, good stability to heat, and shows appreciable change in 10 percent sugar systems. It has a reddish-yellow hue and has good tinctorial strength. It has moderate compatibility with food components and is used in beverages, bakery goods, dessert confections, and ice cream. The common name is sunset yellow FCF.

Fennel   A spice that is the dried, ripe fruit of the herb *Foeniculum vulgare* Mil. It is a seed with licorice flavor. It is used in meat, fish, and sauces as a seasoning.

Fenugreek   The seed, usually in ground form, of the herb *Trigonella foenumgraecum*. It has a maple-like flavor and burnt sugar taste. It is used in curry powder, imitation maple flavor, chutney, and pickles.

Ferric Ammonium Citrate   A nutrient and dietary supplement that is a source of iron, containing 17 percent iron.

Ferric Ammonium Sulfate   A nutrient and dietary supplement that serves as a source of iron.

Ferric Chloride $FeCl_3$. A nutrient and dietary supplement that serves as a source of iron.

Ferric Fructose   A nutrient and dietary supplement that serves as a source of iron.

Ferric Glycerophosphate   A nutrient and dietary supplement that serves as a source of iron.

Ferric Nitrate   A nutrient and dietary supplement that serves as a source of iron.

Ferric Orthophosphate $FePO_4 \cdot xH_2O$. An inert white powder that is a source of iron and produces no discoloration or rancidity. It contains approximately 28 percent iron. It is used as a mineral supplement where rancidity is not a problem. It is used in frozen egg substitute, pasta products, and rice products.

Ferric Oxide Fe_2O_3. A nutrient and dietary supplement that is a source of iron.

Ferric Pyrophosphate   A nutrient and dietary supplement that is a source of iron. It is a tan or yellowish white powder that is insoluble in water but is soluble in mineral acids.

Ferric Sulfate   A nutrient and dietary supplement that is a source of iron.

Ferrous Ammonium Sulfate   A nutrient and dietary supplement that is a source of iron.

Ferrous Carbonate   A nutrient and dietary supplement that is a source of iron.

Ferrous Citrate   A nutrient and dietary supplement that is a source of iron.

Ferrous Fumarate $FeC_4H_2O_4$. A reddish orange to red-brown powder that is a source of iron. It has high bioavailability, and can be used in foods where the red color can be masked. It contains approximately 33 percent iron. It is used as a dietary supplement in breakfast cereals, poultry stuffing, enriched flour, and instant drinks.

Ferrous Gluconate $C_{12}H_{22}FeO_{14} \cdot 2H_2O$. A nutrient and dietary supplement that is a source of iron and a coloring adjunct. It is a yellowish gray to pale greenish yellow powder or granules with a burnt sugar odor. It has a solubility of 1 g in approximately 10 ml of water with slight heating. It is used by the pharmaceutical industry as an iron supplement in vitamin pills. It is used by olive growers to darken the olives to a uniform black color. It can function as an iron fortifier in corn and soy products, breakfast cereals, beverages, and dietary foods.

Ferrous Lactate The ferrous salt of L (+) lactic acid which is used as a source of iron. It is used in foods, such as infant foods, to supplement the needed dietary iron.

Ferrous Sulfate $FeSO_4$. A nutrient and dietary supplement that is a source of iron. It is a white to grayish odorless powder. Ferrous sulfate heptahydrate contains approximately 20 percent iron, while ferrous sulfate dried contains approximately 32 percent iron. It dissolves slowly in water and has high bioavailability. It can cause discoloration and rancidity. It is used for fortification of baking mixes. In the encapsulated form it does not react with lipids in cereal flours. It is used in infant foods, cereals, and pasta products.

Ferrous Tartrate A source of iron.

Flavor The property of a food that results in a sensation of taste in the mouth and odor in the nose. See **Flavors category**.

Flavoring Ingredients that impart or contribute to the flavor of a food. See **Flavors category**.

Flour The food prepared by grinding and bolting cleaned wheat, other than durum wheat and red durum wheat. The baking quality of the flour depends upon the type of wheat, milling process, and treatment applied after milling. Flours classified by process are straight, patent, and clear flour. Flours classified by usage are all-purpose, bread, cake, cracker, and pastry flour. Flours treated after milling include bleached, bromated, enriched bleached, instantized, phosphated, and self-rising flour. Flours from other grains are identified according to the grain source, for example, soy flour. See specific flour.

Folacin See **Folic Acid**.

Folic Acid $C_{19}H_{19}N_7O_6$. A water-soluble B-complex vitamin that aids in the formation of red blood cells, prevents certain anemias, and is essential in normal metabolism. High-temperature processing affects its stability. It is best stored at lower than room temperatures. It is also termed folacin. It is found in liver, nuts, and green vegetables.

Food Starch, Modified See **Modified Starch**.

Fructose $C_6H_{12}O_6$. A sweetener that is a monosaccharide found naturally in fresh fruit and honey. It is obtained by the inversion of sucrose by means of the enzyme invertase and by the isomerization of corn syrup. It is 130 to 180 in sweetness range as compared to sucrose at 100 and is very water soluble. It is used in baked goods because it reacts with amino acids to produce a browning reaction. It is used as a nutritive sweetener in low-calorie beverages. It is also termed levulose and fruit sugar.

Fructose Corn Syrup A sweetener that is an isomerized corn syrup derived from the isomerization of glucose in the syrup to fructose by the enzyme isomerase. Varying levels of fructose syrup are available, being designated 42, 55, and 90 percent fructose. The 42 percent high-fructose corn syrup (HFCS) is a liquid mixture of dextrose, fructose, maltose, isomaltose, and higher saccharides, of which 42 percent is fructose (dry basis). The 55 percent and 90 percent HFCS are liquid mixtures of fructose, dextrose, and higher saccharides containing 55 percent and 90 percent fructose (dry basis), respectively. The range of relative sweetness as compared to sucrose at 100 is 42 percent HFCS: 90 to 95, 55 percent HFCS: 95 to 100, 90 percent HFCS: 100 to 130. HFCS is used in carbonated beverages, canned fruit, frozen desserts, and dairy drinks. It is also termed isomerized syrup, levulose-bearing syrup, and high-fructose corn syrup.

Fruit Sugar See **Fructose**.

Fumaric Acid $C_4H_4O_4$. An acidulant that is a nonhygroscopic, strong acid of poor solubility. It has a solubility of 0.63 g in 100 ml of distilled water at 25°C. It dissolves slowly in cold water, but if mixed with dioctyl sodium sulfosuccinate its solubility improves. The solubility rate also increases with smaller particle size. A quantity of 0.317 kg of fumaric acid can replace 0.453 kg of citric acid. It is used in dry mixes such as desserts, pie fillings, and candy. It is used in dry beverage mixes because it is storage-stable, free-flowing, and nonhygroscopic. It functions as a synergistic antioxidant with BHA and BHT in oil and lard base products. In gelatin desserts, it improves the flavor stability, and increases shelf life and gel strength.

Furcellaran   A gum that is the extract of the red alga *Furcellaria fastigiata*. It swells in cold water and requires heating to 75° to 80°C for solubilization. It forms thermoreversible gels after heating and cooling and has properties between agar and carrageenan. It is also termed Danish agar. It is used in milk puddings, flans, jelly, jam, and gelled meat products.

G

Garlic A spice that is cloves of the herb *Allium sativum*. In its dehydrated form, the flavor enzyme is released only when in combination with water. It exists in powder form and also as salt, chips, and seasoning powder. It is used to flavor meats, vegetables, and sauces.

Garlic Salt A seasoning that is a mix of garlic powder and salt. It is used in sauces and breads.

Gelatin A protein that functions as a gelling agent. It is obtained from collagen derived from beef bones and calf skin (Type B) or pork skin (Type A). Type B is derived from alkali-treated tissue and has an isoelectric point between pH 4.7 and 5.0. Type A is derived from acid-treated tissue and has an isoelectric point between pH 7.0 and 9.0. It forms thermally reversible gels which set at 20°C and melt at 30°C. The gel strength is measured by means of a Bloom Gellometer and ranges from 50 to 300 with a 250 Bloom being the most common. It is used in desserts at 8 to 10 percent of the dry weight, in yogurt at 0.3 to 0.5 percent, in ham coatings at 2 to 3 percent, and in confectionery and capsules at 1.5 to 2.5 percent.

Geranyl Isovalerate A synthetic flavoring agent that is a moderately stable, light yellow liquid of fruity odor. It should be stored in glass or tin containers. It is used in fruit flavors such as apple or pear with applications in beverages, ice cream, candy, and baked goods at 4 to 11 parts per million.

Ghatti   A gum that is a plant exudate obtained from the *Anogeissus latifolia* tree. The gum is formed as a protective sealant when the bark is damaged. It forms viscous mixtures in water at concentrations of 5 percent or greater. Only about 90 percent of the gum is actually soluble in water and has a pH of 4.5. It has similar uses as gum arabic. It is also termed Indian gum. It is used in buttered syrup and as a stabilizer for emulsions.

Ghee   See **Butter Oil**.

Ginger   A spice that is the dried and peeled rhizome of the ginger plant, *Zingiber officinale*. The fragrance ranges from pungent to piquant at once; the flavor can be sharp or cooling depending on the food with which it is used. Fresh (green) ginger is obtained from the cleaned, peeled, and cured rhizome; dried ginger is the fresh product which has been cured and ground for spice. It is used in desserts, meats, sauces, relishes, baked goods, and beverages.

Glacial Acetic Acid $C_2H_4O_2$. An acidulant that is a clear, colorless liquid which has an acid taste when diluted with water. It is 99.5 percent or higher in purity and crystallizes at 17°C. It is used in salad dressings in a diluted form to provide the required acetic acid. It is used as a preservative, acidulant, and flavoring agent. It is also termed acetic acid, glacial.

Gluconic Acid $C_6H_{12}O_7$. An acidulant that is a mild organic acid which is the hydrolyzed form of glucono-delta-lactone. It is prepared by the fermentation of dextrose, whereby the physiological D-form is produced. It is soluble in water with a solubility of 100 g per 100 ml at 20°C. It has a mild taste and at 1 percent has a pH of 2.8. It functions as an antioxidant and enhances the function of other antioxidants. In beverages, syrups, and wine, it can eliminate calcium turbidities. It is used as a leavening component in cake mixes, and as an acid component in dry-mix desserts and dry beverage mixes.

Glucono-Delta-Lactone $C_6H_{10}O_6$. An acidulant, abbreviated GDL. It hydrolyzes to form gluconic acid in water solution and thereby creates the desired pH. The rate of acid formation is affected by temperature, concentration, and the pH of the solution. It has low acid release at room temperature and accelerated conversion into gluconic acid at high temperatures. It is readily soluble with a solubility of 59 g in 100 ml of water at 20°C. It is used in chemical leavening and is used in sausages and frankfurters. It is an acidulant in dessert mixes.

Glucose   See **Dextrose**.

Glucose Syrup   See **Corn Syrup**.

Glutamic Acid $C_5H_9NO_4$. An amino acid that is a white crystalline powder of slight solubility in water. The salt is monosodium glutamate (MSG) which functions as a flavor enhancer in meats. It also is a nutrient, dietary supplement, and salt substitute.

Glutamic Acid Hydrochloride $C_5H_9NO_4 \cdot HCl$. A salt substitute and flavoring agent. It is a white crystalline powder with a solubility in water of 1 g in 3 ml of water.

Gluten   A protein complex formed when water is kneaded with wheat flour which brings about the removal of a large portion of the starch. It forms the elastic framework of dough, entrapping the gas produced by the fermentation of leavening action which results in a risen dough of desired loaf volume and structure. It consists of three water-insoluble proteins — glutenin, gliadin, and mesonin — which form crude gluten when a dough ball is washed with water. Gliadin is of lower molecular weight and provides extensibility as compared to glutenin, which is of higher molecular weight and contributes elasticity. Gluten is available as wheat gluten, corn gluten, and zein. Vital wheat gluten is the most widely used. See **Wheat Gluten**.

Gluten Flour   See **Gluten**.

Glyceral Tribenzoate   A clouding agent used in beverage emulsions.

Glycerin $C_3H_8O_3$. A polyol (polyhydric alcohol) that functions as a humectant, crystallization modifier, and plasticizer. It is a bittersweet liquid which has a high solubility of 71 g per 100 g of water at 25°C. It is a fair oil solvent and has a medium to high hygroscopicity. It is used to maintain a certain moisture content to prevent the drying-out of foods. It is used in confections to maintain the initial level of crystallization of the soft sugar. It also functions as a flavor solvent. Applications include marshmallows, candy, and baked goods.

Glycerol   See **Glycerin**.

Glycerol Dihydroabietate   See **Ester Gum**.

Glycerol Ester of Wood Rosin   See **Ester Gum**.

Glycerol-Lacta-Stearate   See **Glycerol-Lacto-Stearate**.

Glycerol-Lacto-Stearate   An emulsifier that is a glyceryl-lacto ester of fatty acids. It is a monoglyceride esterified with lactic acid which increases the hydrophilicity of the emulsifier. It is used in whipped vegetable toppings, shortenings, cake mixes, and chocolate coating.

Glyceryl Abietate   See **Ester Gum**.

Glyceryl-Lacto Esters of Fatty Acids   Lipophilic emulsifiers that are the lactic acid esters of mono- and diglycerides. They are made by the reaction of mono- and diglycerides or propylene glycol ester with lactic acid, resulting in a compound with more surface activity and slightly more hydrophilicity than the regular mono- and diglycerides. They are used as emulsifiers, plasticizers, and promoters of starch gelatinization. They are used where aeration is required, such as in toppings, cakes, and icings, at levels necessary to obtain the technical effect.

Glyceryl Monolaurate $C_{15}H_{30}O_4$. A monoglyceride emulsifier produced by the esterification of glycerine and lauric acid. It has a melting point of 56°C, a maximum iodine value of 0.5, and a saponification value of 200 to 206. In a highly purified form, it shows antimicrobial properties against microorganisms with the exception of gram-negative organisms. It is effective against gram-negative organisms when formulated with BHA or EDTA. It is used in baked goods, whipped toppings, frosting, glazes, and cheese products.

Glyceryl Triacetate $C_9H_{14}O_6$. A colorless, oily liquid of slight fatty odor and bitter taste. It is soluble with water and is miscible with alcohol and ether. It functions in foods as a humectant and solvent. It is also termed triacetin.

Glycine $C_2H_5NO_2$. A nonessential amino acid that functions as a nutrient and dietary supplement. It has a solubility of 1 g in 4 ml of water and is abundant in collagen. It is used to mask the bitter aftertaste of saccharin, for example, in artificially sweetened soft drinks. It retards rancidity in fat.

Glycoholic Acid $C_{26}H_{43}NO_6$. An emulsifier used in dried egg whites.

Glycyrrhizin   A flavorant and foaming agent derived from the licorice root *Glycyrrhiza glabra*. It is 50 to 100 times as sweet as sugar, is soluble in water, and has a licorice taste. It has foaming and emulsifying properties in water, being used in cocktail mixes and soft drinks. It is used as a flavorant in bacon and imitation whipped products. It is synergistic with sugar, the sweetness being amplified to 100 times that of cane sugar alone. It is used as a sweetener in sugar-free chewing gum and low-fat sugar-free frozen desserts. It is also termed ammoniated glycyrrhizin.

Golden Apple Seed   See **Quince Seed**.

Graham Flour   Another term for whole wheat flour.

Graham's Salt   See **Sodium Polyphosphate** and **Sodium Tetrametaphosphate**.

Grain Sorghum   A thickener that forms a viscous, relatively short opaque paste with typical cereal flavor. It forms gels and has a somewhat higher gelatinization temperature than cornstarch. It is used in wet milling to obtain starch and in dry milling to produce a low-protein flour. It is used where clarity is not required such as in sauces, gravies, and puddings. It is also termed milo.

Grain Vinegar   An acidulant made by the acetous fermentation of dilute distilled alcohol, containing not less than 4 g of acetic acid per 100 ml at 20°C. It is used in mayonnaise, salad dressing, sauces, and catsup. It is also termed distilled vinegar and spirit vinegar.

Granulated Sugar   See **Sugar**.

Grape Color Extract   An aqueous solution of anthocyanin grape pigment made from Concord grapes, or a dehydrated water-soluble powder prepared from the aqueous solution. It contains the common components of the grape juice, but not in the same proportions. It has a red color pigment, with greatest color stability below pH 4.5. The color is stable in the presence of light and some heat. The color intensity increases as the pH declines. It is used at the 0.05 to 0.8 percent range. It may be used for coloring nonbeverage foods.

Grape Seed Oil   The oil obtained from grape seeds which contain an average of 15 percent oil. It is used as a drying oil with seeded raisins to improve their appearance and to prevent sticking. It is also termed raisin seed oil.

Grape Skin Extract   A natural red colorant with a high concentration of red anthocyanic pigments which provide its physicochemical properties. These pigments are responsible for the red, purple, violet, and blue hues of flowers and fruits. It is prepared by aqueous extraction of the fresh seedless marc remaining after the grapes have been pressed in the production of grape juice and wine. It contains the common components of grape juice, but in different proportions. The color depends upon the medium and the pH. In an acid medium and up to pH 4.5 to 5.5, the color is violet and becomes blue at pH 6.5. It has excellent water solubility and fair heat, light, and chemical stability. It can be used in soft drinks at 0.2 to 0.4 percent, in candies at 0.5 to 1.5 percent.

Guaiac Gum   An antioxidant and preservative made from wood resin from the West Indies. It dissolves incompletely but readily in alcohol, ether, chloroform, and in alkali solutions. It is used to retard fat and oil rancidity. It is also termed guaiac resin.

Guaiac Resin   See **Guaiac Gum**.

Guaiacol $C_7H_8O_2$. A precursor of vanillin and santalidol (a synthetic sandalwood fragrance). It is obtained from wood tar by the destructive distillation of hardwood, by the distillation of the phenol fraction of coal tar, or through the use of o-dichlorobenzene. It is processed to yield vanillin.

Guar   A gum that is a galactomannan obtained from the seed kernel of the guar plant *Cyamopsis tetragonoloba*. It is dispersible in cold water to form viscous sols which upon heating will develop additional viscosity. A 1 percent solution has a viscosity range of 2000 to 3500 centipoises. It reacts synergistically with xanthan gum to provide increased viscosity. It is a versatile thickener and stabilizer used in ice cream, baked goods, sauces, and beverages at use levels ranging from 0.1 to 1.0 percent. It is scientifically termed guaran.

Guaran   See **Guar**.

Gum Base   The component of chewing gum that is insoluble in water and remains after chewing. It is prepared by blending and heating several ingredients to include a masticatory substance of vegetable or synthetic origin such as chicle, crown gum, petroleum wax, lanolin, polyethlyene, polyvinyl acetate, or rubber, with a plasticizer such as paraffin and with antioxidants. The gum base is 15 to 30 percent of chewing gum, of which a sweetener is the principal ingredient.

Gum Arabic   See **Ghatti**.

Gum Quince Seed   See **Quince Seed**.

H

Heptyl Cinnamate A synthetic flavoring agent that is a fairly stable, yellow liquid with a hyacinth odor. It should be stored in glass or tin containers. It is used to smooth out fruity flavors and has application in gelatins and puddings at approximately 20 parts per million and in candy, beverages, and ice cream at 2 to 6 parts per million.

Heptyl Formate A synthetic flavoring agent that is a moderately stable, colorless to light yellow liquid of fruity odor. It should be stored in glass or tin containers. It is used in fruit flavors such as apricot, pear, and plum with applications in beverages, ice cream, candy, and baked goods at 1 to 4 parts per million.

Heptyl Isobutyrate A synthetic flavoring agent that is a stable, colorless liquid of fruity odor. It should be stored in glass or tin containers. It is used in flavors for pineapple, pear, and orange with applications in beverages, ice cream, candy, and baked goods at 1 to 3 parts per million.

Heptyl Paraben $C_{14}H_{20}O_3$. A preservative and antimicrobial agent. It is very slightly soluble in water. It may be used in fermented malt beverages to inhibit microbial spoilage and is permitted in beer. It is also termed N-heptyl-para-hydroxybenzoate.

Hesperidin A flavoring agent that is a bioflavonoid found in citrus pulp. It has minor use as a flavorant.

High-Fructose Corn Syrup (HFCS)   A sweetener that is an isomerized corn syrup derived from the isomerization of the glucose in the syrup to fructose by the enzyme isomerase. Varying concentrations of fructose syrup are available, designated 42, 55, and 90 percent fructose. The 42 percent high-fructose corn syrup (HFCS) is a liquid mixture of dectrose, fructose, maltose, isomaltose, and higher saccharides of which 42 percent is fructose, dry basis. The 55 and 90 percent HFCS's are liquid mixtures of fructose, dextrose, and higher saccharides containing 55 and 90 percent fructose, dry basis, respectively. The range of relative sweetness as compared to sucrose at 100 is 42 percent HFCS: 90 to 95, 55 percent HFCS: 95 to 100, 90 percent HFCS: 100 to 130. It is used in carbonated beverages, canned fruit, frozen desserts, and dairy drinks. It is also termed isomerized syrup, levulose-bearing syrup, and fructose corn syrup.

Homogenized Milk   Milk that has been mechanically treated to reduce the size of the fat globules such that after 48 hours of quiescent storage at about 7°C no visible cream separation occurs and the percentage of fat of the upper 100 ml in 946 ml of milk does not differ by more than 10 percent from the fat percentage of the remaining milk. Homogenization makes the milk more homogeneous but also decreases the heat stability of the milk proteins. It is used as a beverage and constituent of other food products. Practically all whole milk sold retail in the United States is homogenized.

Honey   A sweetener that is a natural syrup. It is similar to invert sugar, with a small but variable excess of levulose (fructose). It is formed by the action of the enzyme honey invertase on nectar gathered by bees. The composition and flavor varies with the plant source of the nectar, processing, and storage. A typical composition is 41 percent fructose, 34 percent glucose, 18 percent water, and 2 percent sucrose with a pH of 3.8 to 4.2. It is used in foods as a sweetener.

Horseradish   A spice, the granules obtained from the horseradish plant. The flavor is released with moisture. It has a hot flavor character and has good stability. It is used in sauces.

Hydrated Lime $Ca(OH)_2$. See **Calcium Hydroxide**.

Hydrochloric Acid   HCl. An acid that is the aqueous solution of hydrogen chloride of varying concentrations. It is miscible with water and with alcohol. It is used as an acidulant and neutralizing agent.

Hydrogenated Vegetable Oil Oil that has been hydrogenated to modify the texture from a liquid to a semisolid or solid. The hydrogenization, which is the chemical addition of hydrogen, raises the melting point and converts the oil to a more desirable texture and consistency. It is used in farinaceous foods, confectionary, and desserts.

Hydrolyzed Cereal Solids These are maltodextrins of low DE (Dextrose Equivalent). They function as anticaking agents, bodying agents, carriers, and crystallization inhibitors and are used in dry mixes and desserts.

Hydrolyzed Protein See **Hydrolyzed Vegetable Protein**.

Hydrolyzed Vegetable Protein (HVP) A flavor enhancer obtained from vegetable proteins such as wheat gluten, corn gluten, defatted soy flour, and defatted cottonseed. The proteins are hydrolyzed into their component amino acids after which the reaction mixture is neutralized with sodium carbonate and refined. The refined liquid HVP consists of amino acids, monosodium glutamate, amino acid derivatives, salt, and water. After being stored for several months, the liquid HVP is concentrated into a paste, dried, and ground. A typical dried HVP consists of 40 to 45 percent salt, which is generated during the neutralization process and serves to enhance the mouth feel of the HVP and provide preservation properties. It normally contains 9 to 12 percent monosodium glutamate and the remaining fraction consists of flavor solids. There are two basic types: pale HVP, which functions as a flavor enhancer with delicate spray flavors used in cream-type soups and sauces, and poultry; and dark HVP, which functions as a flavor donor with strong meaty flavors used in stews and broths. HVP is stable under varying processing conditions. It is used to improve flavors in soups, dressings, meats, snack foods, and crackers.

Hydroxylated Lecithin An emulsifier and clouding agent that is a modified crude lecithin of improved water dispersibility. It is manufactured by treating soybean lecithin with peroxide to increase the hydrophilic properties of lecithin. It is partially soluble in water but hydrates readily to form emulsions. It is used in bakery products because it has an apparent synergy with mono- and diglycerides. It is also used in dry-mixed beverages and margarine. It is also termed hydroxylated soybean lecithin.

Hydroxylated Soybean Lecithin See **Hydroxylated Lecithin**.

Hydroxypropyl Cellulose A gum that is nonionic water-soluble cellulose, obtained from the reaction of alkali cellulose with propylene oxide at high temperatures and pressure. It is soluble in water below 40°C, is precipitated as a floc between 40 and 45°C, and is insoluble above 45°C. The precipitation is reversible with the original viscosity being restored upon cooling below 40°C and stirring. It is used in whipped toppings as a stabilizing and foaming aid; in edible food coatings as a glaze and oil/oxygen barrier; and in fabricated foods as a binder. Typical usage level is 0.05 to 1.0 percent.

Hydroxypropyl Methylcellulose A gum formed by the reaction of propylene oxide and methyl chloride with alkali cellulose. It will gel as the temperature is increased in heating and upon cooling will liquefy. The gel temperature ranges from 60 to 90°C, forming semi-firm to mushy gels. It is used in bakery goods, dressings, breaded foods, and salad dressing mix for syneresis control, texture, and to provide hot viscosity. Usage level ranges from 0.05 to 1.0 percent.

Indian Gum   See **Ghatti**.

Indigotine   See **FD&C Blue #2**.

Instantized Flour   A flour made by a milling or agglomerating procedure which makes it readily pourable, providing convenience.

Invert Sugar   A sweetener that is a mixture of equal weights of dextrose (glucose) and levulose (fructose). It is more soluble than sucrose and has higher moisture-retaining properties because of the fructose content. It resists crystallization. It is used in candy and icings because it is sweeter, more soluble, and crystallizes less readily than sucrose.

Invert Sugar Syrup   A sweetener produced by an inversion process. It is produced by solubilizing sucrose in water followed by hydrolization to a mixture of dextrose and fructose using acids, invertase enzyme, or ion exchange resins to catalyze the reaction. Several invert syrups are obtained, such as medium invert consisting of 50 percent sucrose, 25 percent dextrose, 25 percent fructose, and total invert consisting of 3 to 5 percent sucrose, 48 percent dextrose, and 47 percent fructose. It has improved microbiological stability because of its high solids content and it is used in soft drinks. It is also termed sugar syrup, invert.

Iodine   I. A halogen element extracted from Chilean nitrate-bearing earth or from seaweed. It functions by its presence in the thyroid hormones. Iodine deficiency is associated with goiter. Sources are potassium and cuprous iodide and potassium and calcium iodate, of which the iodate form is preferred because of better stability. It is used as a food supplement.

Irish Moss   See **Carrageenan**.

Iron   Fe. A mineral used in food fortification that is necessary for the prevention of anemia, which reduces the hemoglobin concentration and thus the amount of oxygen delivered to the tissues. Sources include ferric ammonium sulfate, chloride, fructose, glycerophosphate, nitrate, phosphate, pyrophosphate and ferrous ammonium sulfate, citrate, sulfate, and sodium iron EDTA. The ferric form (Fe^{+3}) is iron in the highest valence state and the ferrous form (Fe^{+2}) is iron in a lower valence state. The iron source should not discolor or add taste and should be stable. Iron powders produce low discoloration and rancidity. It is used for fortification in flour, baked goods, pasta, and cereal products.

Iron Ammonium Citrate   An anticaking agent used in salt.

Iron Oxide   A trace mineral used as a pigment and colorant. It is used to color pet food.

Iron, Reduced   Iron in a lower valence state, such as the ferrous form (Fe^{+2}). It is used in dry-mix oatmeal.

Isoamyl Acetoacetate   A synthetic flavoring agent that is a stable, colorless liquid of light green leaf-fruity odor. It should be stored in glass or tin containers. It is used in currant and berry flavors for applications in beverages, candy, and ice cream at 5 to 15 parts per million.

Isoamyl Butyrate $C_9H_{18}O_2$. A synthetic flavoring agent that is a stable, colorless liquid of strong fruity odor. It is usually prepared by esterification of isoamyl alcohols with butyric acid. It is soluble in most fixed oils and mineral oil and is insoluble in glycerin and propylene glycol. Storage should be in glass, tin, or resin-lined containers. It is used in fruit flavors such as pineapple, raspberry, and strawberry and has application in dessert gels, puddings, and baked goods at 50 to 60 parts per million.

Isoamyl Formate $C_6H_2O_2$. A synthetic flavoring agent that is a moderately stable, colorless to light yellow liquid of pungent-pear-plum odor, being soluble in most fixed oils, mineral oil, and propylene glycol. Storage should be in a glass or tin container. It is used in fruit flavors such as pear, plum, and peach for application in dessert gels, puddings, candy, and ice cream at 14 to 28 parts per million.

Isoamyl Hexanoate $C_{11}H_{22}O_2$. A synthetic flavoring agent that is a stable, colorless liquid of fruity odor. It is soluble in alcohol, fixed oils, and mineral oil. Storage should be in glass, tin, or resin-lined containers. It is used in fruit flavors such as banana and pineapple for applications in desserts, candy, and ice cream at 4 to 22 parts per million.

Isoascorbic Acid See **Ascorbic Acid**.

Isobutyl Cinnamate $C_{13}H_{16}O_2$. A synthetic flavoring agent that is a stable, colorless to light yellow liquid of fruity odor. It is miscible with alcohol, chloroform, and ether but is practically insoluble in water. Storage should be in glass or tin-lined containers. It is used in fruit flavors such as cherry and prune with applications in beverages, ice cream, candy, and baked goods at 1 to 5 parts per million.

Isobutyl Formate A synthetic flavoring agent that is a stable, colorless liquid of fruity odor. Storage should be in glass or tin containers. It is used in fruit flavors such as pear, raspberry, and other berry flavors with applications in beverages, ice cream, candy, and baked goods at 2 to 18 parts per million.

Isolated Soy Protein See **Soybean Protein Isolate**.

Isomerized Syrup See **High-Fructose Corn Syrup**.

Isopropyl Citrate An antioxidant that reacts with metal ions that might catalyze oxidative reactions and thus will prevent rancidity. It is made by reacting citric acid (not soluble in fats and oils) with isopropyl alcohol which readily dissolves in oil and thus enables the citrate to dissolve in oil. It is used in vegetable oils.

Karaya A gum, the dried exudate from the *Sterculia urens* tree which is native to India. It does not dissolve in water but swells to form a colloidal sol with a rate of hydration depending on mesh size. A 3 to 4 percent sol will result in a heavy gel and for higher concentrations the gum must be cooked under steam pressure to solubilize. It has a pH of 4.5 to 4.7. It functions as a binder and adhesive. It is used in baked goods, denture adhesives, toppings, and frozen desserts.

Kelp A brown seaweed that grows in ocean water. The principal commercial species include *Macrocystis pyrifera* and *Laminaria hyperboria*. It is a source of alginic acid, which is used to produce alginate gum which functions as a water control agent. It contains the trace minerals potassium, sodium, calcium, and iodine. It is used as a source of iodine, as a flavor enhancer, as a nutrient and dietary supplement, and as a source of alginates.

Kola Nut The seed of *Cola nitida* or other *Cola* species. The nut contains approximately 1.5 percent caffeine and is used in beverages and as an adjunct with other flavors.

Lactalbumin A milk protein obtained from whey by acidifying to pH 5.2, the isoelectric point, followed by coagulation by heat. It is not coagulated by rennin as is casein and is nonfunctional in its properties. It is used for nutritional purposes as a source of protein. It is used in cereals and breads where its relative inertness minimizes complications caused by other milk proteins during baking. It is also termed milk albuminate.

Lactic Acid $C_3H_6O_3$. An acidulant that is a natural organic acid present in milk, meat, and beer, but is normally associated with milk. It is a syrupy liquid available as 50 and 88 percent aqueous solutions, and is miscible in water and alcohol. It is heat stable, nonvolatile, and has a smooth, milk acid taste. It functions as a flavor agent, preservative, and acidity adjuster in foods. It is used in Spanish olives to prevent spoilage and provide flavor, in dry egg powder to improve dispersion and whipping properties, in cheese spreads, and in salad dressing mixes.

Lacto Esters of Propylene Glycol See **Lactylated Fatty Acid Esters of Glycerol and Propylene Glycol**.

Lactoglobulin A protein that is a complex of closely related proteins known as beta-globulins obtained from the whey fraction of milk. It is crystallizable and heat-denaturable.

Lactose $C_{12}H_{22}O_{11}$. A disaccharide carbohydrate that occurs in mammalian milk except that of the whale and the hippopotamus. It is principally obtained as a cows' milk derivative. It is also termed milk sugar and it is a reducing sugar consisting of glucose and galactose. Its most common commercial form is alpha-monohydrate, with the beta-anhydride form available to a lesser extent. All forms in solution will equilibrate to a beta:alpha ratio of 62.25:37.75 at 0°C. It is about one-sixth as sweet as sugar and is less soluble. It functions as a flow agent, humectant, crystallization control agent, and sweetener. It is used in baked goods for flavor, browning, and tenderizing and in dry mixes as an anticaking agent.

Lactylated Fatty Acid Esters of Glycerol and Propylene Glycol An emulsifier made by the reaction of a propylene glycol ester with lactic acid. It has more surface activity and is slightly more hydrophilic than mono- and diglycerides. It is used mainly where aeration is required, such as in toppings, cake mixes, and icings. It is used at levels required to produce the intended effect, such as 0.6 percent in fluid whipped topping and 0.5 percent in coffee whitener.

Lactylic Esters of Fatty Acids An emulsifier that is mixed fatty acid esters of lactic acid and its polymers. It is dispersible in water and soluble in organic solvents and vegetable oils. It functions as a foaming agent in starch/protein systems and is used in puddings and coffee whiteners.

Larch See **Arabinogalactan**.

Lard A fat rendered from hogs, consisting principally of oleic and palmitic fatty acids. It has a Wiley melting point of 88 to 110°F. It is rapidly chilled, resulting in an opaque, firm consistency rather than a translucent, greasy appearance. It is used in cake mix.

Lauric Acid $C_{12}H_{24}O_2$. A fatty acid obtained from coconut oil and other vegetable fats. It is practically insoluble in water but is soluble in alcohol, chloroform, and ether. It functions as a lubricant, binder, and defoaming agent.

Leavening Agents Acidic agents that chemically react with alkaline sodium bicarbonate to produce carbon dioxide gas. This reaction is initiated by moisture and completed by heat as the prepared mixture is baked. The value of the leavening agent relates to the rate upon which carbon dioxide is released from sodium bicarbonate, the suitability of the release rate to the product, and the mixing-raising-baking cycle. Leavening agents include tartaric acid, monocalcium phosphate, sodium acid pyrophosphate, sodium aluminum phosphate, and acidic acid.

Lecithin   An emulsifier that is a mixture of phosphatides which are typically surface-active. It is now commercially obtained from soybeans; previously it was obtained from egg yolk. It is used in margarine as an emulsifier and antispatter agent; in chocolate manufacture it controls flow properties by reducing viscosity and reducing the cocoa butter content from 3 to 5 percent; it is used as a wetting agent in cocoa powder, fillings, and beverage powders; an antisticking agent in griddling fat; and in baked goods to assist the shortening mix with other dough ingredients and to stabilize air cells. Typical usage levels range from 0.1 to 1.0 percent.

Lecithinated Soy Flour   Soy flour to which lecithin is added. The lecithin contributes emulsification and pan release properties. It is used in breading, caked foods, and dough mixes.

Lemon Oil   A flavoring agent that is the oil obtained from lemon fruit. It is used to impart lemon flavor and is used in reconstituted lemon juice.

Levulose   See **Fructose**.

Levulose-Bearing Syrup   See **High-Fructose Corn Syrup**.

Licorice   A flavoring agent made from dried root portions of *Glycyrrhiza glabra*. The obtainable forms are licorice root, licorice extract powder, and licorice extract. The extract is used in candy, baked goods, and beverages; the major licorice use is in tobacco.

Lime   See **Calcium Oxide**.

Limestone   See **Calcium Carbonate**.

Limestone, Dolomitic   A type of limestone containing magnesium carbonate.

Limonene $C_{10}H_{16}$. An antioxidant and flavoring agent that occurs in lemons, oranges, and pineapple juice, being obtained from the oils. It is a colorless liquid which is insoluble in water and propylene glycol, very slightly soluble in glycerine, and miscible with alcohol, most fixed oils, and mineral oil. It prevents or delays enzymatic browning-type oxidation.

Locust Bean Gum   A gum that is a galactomannan obtained from the plant seed from the locust bean tree known as *Ceratonia siligua*. Its properties include swelling partially in cold water but requiring heating to approximately 82°C for complete solubility. It provides high viscosity, forms gels with xanthan gum upon heating and cooling of the solution, and functions as a water binder. It can make agar or carrageenan gels more elastic. Its uses include processed cheese, ice cream, bakery products, soups, and pies. Typical usage level is 0.1 to 1.0 percent. It is also called carob gum, Saint John's bread, and is scientifically called carubin.

Low-Methoxyl Pectin   A gum derived from pectinic acid. It differs from pectin in having a lower degree of methylation, less than 50 percent. It is also not as sensitive to pH and does not require sugar for gel formation. It forms thermally reversible gels with calcium salts and boiling may be required for solubility if the methoxyl content is low. It is used in low-calorie jellies at levels of 0.8 to 1.4 percent, in dessert and bakery jellies at levels of 1 to 1.5 percent, and in fruit gels at levels of 0.5 to 1.4 percent.

Lycasin   A sweetener produced by the hydrogenation of glucose syrup. It is one-fourth to one-half as sweet as sucrose.

Mace   A spice that is the aril or skin covering of the nutmeg *Myristica fragrans* Houtt. It is more pungent in flavor than nutmeg. The whole mace is used in cooked fruit, pickles, and preserves, while ground mace is used in breads, cakes, chocolate pudding, and fruit salad.

Magnesium   Mg. A metallic element that is involved in certain bodily functions. Sources of magnesium include magnesium chloride and magnesium oxide. It functions as a nutrient and dietary supplement.

Magnesium Carbonate $MgCO_3$. An anticaking agent and general purpose food additive. It is practically insoluble in water but is more soluble in water containing carbon dioxide. It imparts a slightly alkaline reaction to the water. It is used as an alkali in sour cream, butter, and canned peas. It is used as an anticaking agent in table salt and dry mixes. It assists in providing clarity in algin gels and functions as a filler in dental impression materials.

Magnesium Caseinate   The magnesium form of caseinate which is marginal in functionality as compared to other forms of caseinates. It can be used in bakery goods, drinks, and dietary applications. See **Caseinates**.

Magnesium Chloride $MgCl_2$. A source of magnesium, a color-retention agent, and firming agent. It exists as colorless flakes or crystals and is very soluble in water.

Magnesium Hydroxide $Mg(OH)_2$. An alkali that is a general purpose food additive. It exists as a white powder and has poor solubility in water and in alcohol. In frozen desserts it will increase the tendency for fat globules to clump, which results in an increase in dryness. It reacts with triglycerides in fatty acids to form soaps. It also functions as a drying agent in foods.

Magnesium Laurate   The magnesium salt of lauric acid which functions as a binder, emulsifier, and anticaking agent.

Magnesium Myristate   The magnesium salt of myristic acid which functions as a binder, emulsifier, and anticaking agent.

Magnesium Oleate   The magnesium salt of oleic acid which functions as a binder, emulsifier, and anticaking agent.

Magnesium Oxide   MgO. A source of magnesium which functions as a nutrient and dietary supplement. It exists as a bulky white powder termed light magnesium oxide or as a dense white powder known as heavy magnesium oxide. It is practically insoluble in water and is insoluble in alcohol.

Magnesium Palmitate   The magnesium salt of palmitic acid which functions as a binder, emulsifier, and anticaking agent.

Magnesium Silicate   A white powder that is insoluble in water and functions as an anticaking agent. It is used in salt. It also is a processing aid and adsorbent which functions as an anticaking agent and remover of undesirable proteins during filtration. It is insoluble and a 10 percent slurry has a pH of approximately 7.0. It aids in the processing of beverages, food products, and pharmaceuticals by removing protein/tannin complex constituents through surface area and adsorptive effects.

Magnesium Stearate   The magnesium salt of stearic acid which functions as a lubricant, binder, emulsifier, and anticaking agent. It is a white powder that is insoluble in water. It is used as a lubricant or die release in tableting pressed candies and is also used in sugarless gum and mints.

Magnesium Sulfate $MgSO_4$. A nutrient and dietary supplement which is also termed Epsom salt. It is soluble in water and its solutions are neutral. It exists as crystals with a cooling, saline, bitter taste.

Maize Meal The meal obtained by grinding maize (Indian corn).

Maize Starch See **Cornstarch**.

Maleic Acid $C_4H_4O_4$. A white crystalline powder that functions as a preservative. It is moderately toxic and is a strong irritant to living tissue. It is used in pharmaceuticals for the preparation of maleate salts of antihistamines. It is used in foods as a preservative for fats and oils.

Malic Acid $C_4H_6O_5$. An acidulant that is the predominant acid in apples. It exists as white crystalline powder or granules and is considered hygroscopic. As compared to citric acid, it is slightly less soluble but is still readily soluble in water with a solubility of 132 g per 100 ml at 20°C. It has a stronger apparent acid taste and has a longer taste retention than citric acid which peaks faster but does not mask the aftertaste as effectively. A quantity of 0.362 to 0.408 kg of malic acid is equivalent to 0.453 kg of citric acid and to 0.272 to 0.317 kg of fumaric acid in tartness. At temperatures above 150°C it begins to lose water very slowly to yield fumaric acid. It is used in soft drinks, dry-mix beverages, puddings, jellies, and fruit filling. It is used in hard candies because it has a lower melting point (129°C) than citric acid which improves the ease of incorporation.

Malt A source of the enzyme alpha-amylase which hydrolyzes starch to fermentable sugars such as dextrins and maltose. It is produced by the controlled sprouting of grains, usually barley, followed by drying to produce three general classes of malt differing in amylase content. These classes are brewer's malt, distiller's malt, and gibberellin malt. Malt is used in the brewing industry and as a supplement to flour to increase the alpha-amylase content.

Malted Barley The barley produced under the controlled sprouting of the barley grain followed by drying to obtain the formation of high levels of alpha-amylase and some increase in beta-amylase. These enzymes hydrolyze starch to dextrins and maltose. There are three general classes of malt: brewer's malt, distiller's malt, and gibberellin malt. It is principally used in the brewing industry. In doughs, the malt increases the fermentation rate and improves baking properties.

Malted Cereal Syrup The syrup obtained from barley and other grains, as differentiated from malt syrup which is obtained only from barley. It is used to contribute flavor in baked goods and is a nutrient in yeast fermentation. It is also termed extract of malted barley and corn.

Malt Extract A flavorant formed by extracting the water-soluble enzymes from barley and evaporating to form a concentrate that contains D-alpha-amylase enzyme. This enzyme hydrolyzes starch to dextrin and maltose. It is used to provide malt flavor, and in conjunction with spices, seasonings, and flavors.

Malt Flour The flour prepared by the drying and grinding of barley or wheat sprouted under controlled conditions. It can be used as a malt supplement or converted to malt syrups. The malt functions to modify starch during initial stages of baking.

Maltodextrin The product obtained from the partial acid or enzymatic hydrolysis of starch, in the same manner as corn syrup except the conversion process is stopped at an earlier stage. It has a dextrose equivalent of less than 20 and basically is not sweet and is not fermentable. It has fair solubility. It functions as a bodying agent, bulking agent, texturizer, carrier, and crystallization inhibitor. It is used in crackers, puddings, and candies.

Maltol $C_6H_6O_3$. A flavor enhancer used as a synthetic flavoring substance, the function of which is related to ethyl maltol. It occurs naturally in chicory, cocoa, coffee, and cereals. It does not contribute a flavor of its own, but modifies the inherent flavors. As compared to ethyl maltol, it is one-half to one-sixth as effective. It is less soluble, having a solubility of 1 g in 82 ml of water at 25°C. It has a melting range of 160 to 164°C. It is used to enhance the flavor and aroma of fruit, vanilla, and chocolate-flavored foods and beverages. It is also used in beverages and desserts with a typical usage range of 10 to 200 parts per million.

Maltose $C_{12}H_{22}O_{11}$. A sweetener formed by the enzymatic action of yeast on starch. It consists of two dextrose molecules. Maltose dissolves and crystallizes slowly in aqueous solutions, and is less sweet and more stable than sucrose. It is used in combination with dextrose in bread and in instant foods, and is also used in pancake syrups.

Malt Syrup The syrup obtained from barley by extraction and evaporation of the worts to 80 to 81 percent solids. It is used as a malt flavor component, source of malt and protein, and to provide color. It is used in bakery goods such as rolls and bagels at 1 to 3 percent of the flour weight, in soybean milk at 3 to 7 percent, and in malt base at 1 to 3 percent.

Malt Vinegar   A vinegar made by the alcoholic and subsequent acetous fermentation of an infusion of malted barley and/or cereals or a concentrate thereof, which has been enzymatically converted by the malting process. It contains a minimum of 4 percent acid content expressed as acetic acid and is brown to dark brown in color. It functions as an acidulant and preservative in foods.

Manganese   Mn. A metallic element that functions as a nutrient and dietary supplement. It is necessary for normal bone and tendon structure, central nervous system functionality, and glucose utilization. Sources include manganese carbonate, chloride, oxides, and sulfates. These sources differ in solubility.

Manganese Carbonate   A source of manganese that functions as a nutrient and dietary supplement. See **Manganese**.

Manganese Chloride $MnCl_2 \cdot 4H_2O$. A source of manganese that functions as a nutrient and dietary supplement. It exists as crystals which are readily soluble in room temperature (22°C) water and are very soluble in hot water. See **Manganese**.

Manganese Oxide   MnO. A source of manganese that functions as a nutrient and dietary supplement. See **Manganese**.

Manganese Sulfate $MnSO_4 \cdot H_2O$. A source of manganese that functions as a nutrient and dietary supplement. It exists as a powder which is readily soluble in water. See **Manganese**.

Mannitol $C_6H_{14}O_6$. A polyol (polyhydric alcohol) that functions as a sweetener, humectant, and bulking agent. It has low hygroscopicity and poor oil solvency. It is approximately 22 percent soluble in water and is approximately 72 percent as sweet as sugar, exhibiting a cool, sweet taste. It functions as a dusting agent with starch in chewing gum. It is used in sugarless candy, chewing gum, cereal, and pressed mints.

Maple Sugar   A sweetener obtained by concentrating the sap of the maple sugar tree. It consists of approximately 95 percent sucrose, 2 percent invert sugar, and ash. This is the dry form of maple syrup which is used in syrups and candy.

Maple Syrup   A sweetener made by concentrating the sap of the sugar maple tree by boiling at atmospheric pressure. The characteristic color and flavor are developed by heating above 100°C. The concentration at reduced pressure or by freeze-drying gives a sweet, colorless syrup. The characteristic flavor is derived from the volatile oil in the sap. On a dry basis it is approximately 95 percent sucrose, 2 percent invert sugar, and ash. It is used in syrups and candies.

Margarine   A product whose consistency and application are similar to those of butter. It is made by emulsifying vegetable oils with milk, followed by crystallization and kneading. Vegetable oils or mixtures of vegetable oils and animal fat may be used. It contains not less than 80 percent fat and is also termed oleomargarine. It is used as a spread and as a source of fat for baked goods and desserts.

Marjoram   A spice that is the dried leaves of the herb *Majorana hortensis* Moench. It has a mellow flavor and is distinctively aromatic. The flavor is close to that of oregano. Marjoram is used in soups, sauces, meats, and fish.

Methylcellulose   A gum composed of cellulose in which the methoxyl groups replace the hydroxyl groups. It is soluble in cold water but insoluble in hot water. Solutions increase in viscosity upon heating, gel at 50 to 55°C, and liquefy upon cooling. It is used in baked goods for moisture retention, and in fruit pie fillings for the reduction of water absorption into the pie crust during baking. It is also used in breaded shrimp where it functions to form an oil barrier film.

Methylethylcellulose   A gum that is the methyl ether of ethyl cellulose. It is soluble in water below 60°C, and insoluble above 60°C, with the gum precipitating as a swollen floc which redissolves upon cooling the solution below 60°C. The solutions have a pH of 5 to 7 and are pseudoplastic, decreasing in viscosity with increase in shear. Its application usage is based on its whipping properties and ability to form stiff foams. It is used in whipped toppings at 0.5 to 0.8 percent, in meringues at 1 percent, and in other aerated confectionary products.

Methyl p-Hydroxybenzoate   See **Parabens**.

Methyl 2-Methylthiopropionate   A synthetic flavoring agent that is a colorless liquid of slightly fruity odor with a suggestion of sulfur. It polymerizes slowly and should be stored in glass or tin containers. It is used in pineapple flavors to give the true note of pineapple. It has applications in beverages, ice cream, candy, and baked goods at 0.5 to 1 part per million.

Methylparaben $C_8H_8O_3$. An antimicrobial agent which is a white free-flowing powder. It is active against yeast and molds over a wide pH range. See **Paraben**.

Methyl Polysilicone   See **Dimethylpolysiloxane**.

Methyl Salicylate $C_8H_8O_3$. A synthetic flavoring agent that is a fairly stable, colorless to light yellow liquid of wintergreen odor. It should be stored in glass or tin-lined containers. It is soluble in alcohol and glacial acetic acid and very slightly soluble in water. It is used in chewing gum at 8400 parts per million, in candy at 850 parts per million, and in beverages and baked goods at 50 to 60 parts per million. It is also termed wintergreen oil.

Methyl Silicone   See **Dimethylpolysiloxane**.

3-Methylthiopropionaldehyde   A synthetic flavoring agent that is a colorless to light yellow liquid with an intense meat odor. It polymerizes with age and is stable in a 50 percent alcohol solution. It should be stored in glass containers. It is used at low concentrations for meat and broth flavors for applications in meats and condiments at 3 parts per million and in baked goods and beverages at 0.5 part per million.

Microcrystalline Cellulose   A gum that is the nonfibrous form of cellulose, an alpha-cellulose. It is dispersible in water but not soluble, requiring considerable energy to disperse and hydrate. In this form it is used in dry applications such as tableting, capsules, and shredded cheese where it functions as a non-nutritive filler, binder, flow aid, and anticaking agent. By addition of carboxymethylcellulose to the alpha-cellulose prior to drying, improved functional properties of hydration and dispersion are obtained. This product is designed for use in water dispersions, being insoluble in water but dispersing in water to form colloidal sols below 1 percent and white opaque gels above the 1 percent usage level. It is used as a heat shock stabilizer and bodying agent in frozen desserts, as an opacifier in low-fat dressings, as a foam stabilizer in whipped toppings, and as an emulsifier in dressings.

Milk   The natural secretion of the mammary glands of female mammals for the feeding of their young. It is commercially considered here as cows' milk which consists, on the average, of 3.5 percent fat, 5 percent lactose, 3.5 percent protein, and 0.7 percent ash. It has a bland, slightly sweet flavor, a yellowish-white color, and a specific gravity of 1.032. It functions as a base for ice cream, yogurt, beverages, and cheese. It is also the source of skim milk, cream, whey, casein, lactose, and milk solids–not-fat.

Milk Albuminate   See **Lactalbumin**.

Milk Chocolate   See **Chocolate**.

Milkfat   The fat of milk which exists in milk as an emulsion of small fat globules in an aqueous phase. It is the only fat in which butyric acid is a component of the glycerides. It has a delicate and pleasant flavor. Approximately 95 percent of the total milk lipids are triglycerides. The average fat content of milk is 3.5 to 3.8 percent. It is used as a source of fat in bakery products, confections, and frozen desserts. It is also termed butter fat.

Milkfat, Anhydrous   See **Butter Oil**.

Milk Powder   The dry, whole milk that is produced by a spray- or roller-drying process to remove the water fraction. The dry form offers convenience of transportation, utility, and stability. It is used in soup mixes and dessert mixes.

Milk Solids–Not-Fat   The dry form of skim milk. It contains not more than 1.5 percent fat and not more than 5 percent moisture. It has excellent flavor, nutritional value, and functional properties such as water binding, emulsification, and foam formation. It is used in ice cream mix, baked goods, and desserts. It is also termed nonfat dry milk, skim milk powder, and dried skim milk.

Milk Sugar   See **Lactose**.

Milo   See **Grain Sorghum**.

Mint   A spice derived from any one of the plants of the mint family (Labiatae) of which there are numerous varieties. Only spearmint and peppermint are commercially important. Mint is used in mint sauce, fruit cocktails, and beverages in either its dried or fresh forms.

Modified Food Starch   See **Modified Starch**.

Modified Starch   The product resulting from the treatment of starch with certain chemicals to modify the physical characteristics of the native starch. This produces more desirable or useful characteristics such as improved solubility, acid stability, and texture. It is used in desserts, pie fillings, sauces, gravies, and fabricated foods as a thickener, binder, and stabilizer.

Molasses   The by-product of the manufacture of sugar from sugar cane in which the syrup is separated from the crystals. The highest grade is edible molasses which is most often found as table syrup. The lowest grade is blackstrap molasses. Molasses is a strongly flavored, dark colored syrup containing 70 to 80 percent solids of which 50 to 75 percent is sucrose and invert sugar. It is used in syrups and in the production of caramel.

Monoammonium Glutamate $C_5H_{12}N_2O_4 \cdot H_2O$. A flavor enhancer that is a white crystalline material obtained from glutamic acid. It is soluble in water, having a solubility of 90 g per 100 ml of water at 20°C. It functions similarly to monosodium glutamate. It is used as a flavor enhancer in low-salt diets because of its negligible sodium content.

Monoammonium Glycyrrhizinate See **Ammoniated Glycyrrhizin**.

Monoammonium Phosphate $NH_4H_2PO_4$. An acidulant. A fast-reacting leavening agent suggested for use in items baked at high temperatures. It is also used as a nutrient for yeast in bread manufacture, supplying nitrogen and phosphorus.

Mono- and Diglycerides   A lipophilic emulsifier that consists of both monoglycerides and diglycerides. It is made by reacting glycerol with specific fats or oils. The consistency varies from yellow liquid to ivory-colored plastic to hard solids of bland odor and taste. It is the most commonly used emulsifier in foods. It is used in numerous applications for example in baked goods, frozen desserts, whipped toppings, and margarine for a variety of functions. Typical usage levels range from 0.05 to 0.50 percent.

Monocalcium Phosphate $Ca(H_2PO_4)_2$. An acidulant, leavening agent, and nutritional supplement that exists as white crystals or granular powder. It is sparingly soluble in water. It is used as an acidulant in breads and dry-mix beverages; as a source of calcium in fruit jellies, preserves, and cereals; and as a component of dough conditioners. It is also of restricted use as a chemical leavening agent because it releases about 67 percent of the carbon dioxide during the initial mixing and this is generally too rapid. It is also termed calcium phosphate monobasic, calcium biphosphate, and acid calcium phosphate.

Monoglyceride   A lipophilic emulsifier prepared by the direct esterification of fatty acids with glycerol or by the interesterification between glycerol and other triglycerides. It is insoluble in water. It provides emulsion stability, prevents fat separation, and also functions as a foaming agent, defoaming agent, and dispersant. It is most often used in combination with a diglyceride. It is used in ice cream, peanut butter, puddings, and numerous other applications. The typical usage level is 0.05 to 0.40 percent.

Monoglyceride Citrate   A sequestrant that is a mixture of glyceryl mono-oleate and its citric acid mono-ester. It is soluble in common fat solvents and alcohol and is insoluble in water. It is used in antioxidant formulations for addition to fats and oils at less than 200 parts per million. It functions as an antioxidant synergist in peanut oil at a maximum usage level of 100 mg/kg individually or in combination. In margarine, it is used at no more than 0.01 percent.

Monoglyceride, Distilled   See **Distilled Monoglyceride**.

Monoisopropyl Citrate   A sequestrant used in fats and oils.

Monopotassium Monophosphate   See **Monopotassium Phosphate**.

Monopotassium Phosphate KH_2PO_4. A buffer, neutralizing agent, and sequestrant. It is mildly acid, with a pH of 4.5, and fairly soluble in water, with a solubility of 25 g in 100 ml at 25°C. It is used in whole eggs for color preservation and is used in low-sodium products, milk, and meat products. Typical usage ranges from 0.1 to 0.5 percent. It is also termed potassium dihydrogen orthophosphate, potassium phosphate monobasic, and monopotassium monophosphate.

Monosodium Dihydrogen Orthophosphate   See **Monosodium Phosphate**.

Monosodium Glutamate (MSG) $C_5H_8NNaO_4 \cdot H_2O$. A flavor enhancer that is the sodium salt of glutamic acid, an amino acid. It is a white crystal that is readily soluble in water. It intensifies and enhances flavor but does not contribute a flavor of its own. It may be present as one of the amino acids or in a free form, which is how it effectively enhances the flavor of foods. It is produced through a fermentation process of molasses. It is used at 0.1 to 1.0 percent in meats, soups, and sauces.

Monosodium Monophosphate See **Monosodium Phosphate**.

Monosodium Phosphate NaH_2PO_4. An acidulant, buffer, and seques-trant that is mildly acid, with a pH of 4.5 and very soluble in water, with a solubility of 87 g per 100 ml of water at 25°C. It is used as an acidulant in effervescent powders and laxatives. It is also used in soft drink dry-mix formulations, in cheese, and in carbonated beverages. It is also termed monosodium dihydrogen orthophosphate; sodium phosphate, monobasic; sodium biphosphate; and monosodium monophosphate.

Monosodium Phosphate Derivatives of Mono- and Diglycerides See **Mono- and Diglycerides**.

Mono-Tertiary-Butylquinone See **Tertiary Butylhydroquinone**.

Mustard A flavorant made from the dried, ripe seed of several closely related genera, species, and varieties of the family Cruciferae; the seed of a plant of the cabbage family. It is used as a flavorant in baked goods, sauces, salad dressings. It also functions as an emulsifier in salad dressings. The ground seed is used for spices.

Mustard Flour The ground seed of the mustard plant from which some of the oil and most of the hulls have been removed. It is used in salad dressings and sauces, and as a condiment.

Mustard Oil See **Allyl Isothiocyanate**.

Mustard Seed A spice of which there are several varieties, the dry mustards being of the hot or mild type. It is used in meats, sauces, and salad dressings.

Myristic Acid $C_{14}H_{28}O_2$. A fatty acid obtained from coconut oil and other fats. It has poor water solubility but is soluble in alcohol, chloro-form, and ether. It is used as a lubricant, binder, and defoaming agent.

Natamycin $C_{33}H_{47}NO_{13}$. A preservative for use as a coating on the surface of Italian cheeses to prevent the growth of mold or yeast. It is tasteless, odorless, colorless, and does not penetrate the cheese. It is very active against virtually all molds and yeasts, but does not affect bacteria, thus not affecting the ripening and flavor improvement process of cheese. It can be applied as a dip, spray, or by other methods such as incorporation into the cheese coatings. It is used at levels ranging from 300 to 2000 parts per million. It is also termed pimaricin.

Natural Sugar See **Turbinado Sugar**.

Niacin $C_6H_5NO_2$. A water-soluble B-complex vitamin that is necessary for the growth and health of tissues. It prevents pellagra. It has a solubility of 1 g in 60 ml of water and is readily soluble in boiling water. It is relatively stable in storage and no loss occurs in ordinary cooking. Sources include liver, peas, and fish. It was originally termed nicotinic acid and also functions as a nutrient and dietary supplement.

Niacinamide $C_6H_6N_2O$. A nutrient and dietary supplement that is an available form of niacin. Nicotinic acid is pyridine beta-carboxylic acid and nicotinamide, which is another term for niacinamide, is the corresponding amide. It is a powder of good water solubility, having a solubility of 1 g in 1 ml of water. Unlike niacin, it has a bitter taste.

Nicotinamide See **Niacinamide**.

93

Nicotinic Acid   See **Niacin**.

Nitrate   The salt of nitric acid. It is used in meat curing to develop and stabilize the pink color associated with cured meat. By itself, it is not effective in producing the curing reaction until it is chemically reduced to nitrite. It has an effect on flavor and also functions as an antioxidant. It is available as sodium and potassium nitrate, with the sodium form being more common.

Nitrite   The salt of nitrous acid. It is used in meat curing to develop and stabilize the pink color associated with a cured meat and to affect flavor and function as an antioxidant. Nitrites convert to nitric oxide, which reacts with the myoglobin pigments (purple-red) to form nitrosomyoglobin (dark red). Nitrosomyoglobin plus heating to 130 to 140°F results in the formation of the stable pigment nitrosohemochrome, resulting in the cured meat color. It has bacteriostatic properties as an inhibitor of *Clostridia* organisms, especially *Clostridium botulinum*, and, therefore, nonsterile canned hams can be produced. Sources are sodium and potassium nitrite, with the sodium form being more commonly used.

Nitrous Oxide N_2O. A noncombustible gas used as a propellant in certain dairy and vegetable fat whipped toppings contained in pressurized containers.

(Gamma)-Nonalactone $C_9H_{16}O_2$. A synthetic flavoring agent that is a colorless to yellow liquid of strong, coconut-like odor. It is soluble in most fixed oils, mineral oil, and propylene glycol. It is stable in acids and unstable in alkali and should be stored in glass, tin, or aluminum containers. It is used in coconut flavors and has application in gelatins, puddings, baked goods, candy, and ice cream at 11 to 55 parts per million. It is also termed aldehyde C-18.

Nonfat Dry Milk   See **Milk Solids – Not-Fat**.

Norbixin   See **Annatto**.

Nordihydroguaiaretic Acid (NDGA)   An antioxidant that has poor solubility and shows evidence of discoloration in the presence of metal salts. It is used to a limited extent to retard rancidity.

Nutmeg   A spice obtained from the nutmeg tree *Myristica fragrans*. It is related to mace, which is obtained from the covering of nutmeg. Nutmeg is used in eggnog, cakes, fruit, and puddings.

Oat A grain that is a source of oat flour. It is used in porridge, grits, and oatmeal.

Oat Flour Fine-mesh ground oats with the hull removed. It has some antioxidant properties and is blended with other flours to retard rancidity.

Oatmeal The food produced by grinding oats after removal of the husk.

Octafluorocyclobutane A propellant and aerating agent for foamed or sprayed foods.

(Gamma)-Octalactone A synthetic flavoring agent that is a stable, colorless to slightly yellow liquid of peach odor. It should be stored in glass or tin containers. It is used in flavors for peach with applications in baked goods, candy, and ice cream at 5 to 17 parts per million.

1-Octanol $C_8H_{16}O$. A synthetic flavoring agent that is a colorless, stable liquid of sharp fatty odor. It is soluble in alcohol, most fixed oils, mineral oil, and propylene glycol. It should be stored in glass or tin containers. It is used in essential oils for application in beverages, candy, and baked goods at 1 to 3 parts per million. It is also termed octyl alcohol.

Octyl Alcohol See **1-Octanol**.

Oleic Acid $C_{18}H_{34}O_2$. An unsaturated fatty acid that functions as a lubricant, binder, and defoamer.

Oleomargarine   See **Margarine**.

Oleoresins   Solvent-free extractions from spices that contain the volatile and nonvolatile flavor components. They are the closest replacements for a spice, and are used in seasonings for a variety of foods.

Oleoresin Paprika   A seasoning and colorant that is the solvent-free extraction containing the volatile and nonvolatile flavor components of paprika. It is the closest replacement for paprika. As a colorant, the pigment is a red-orange carotenoid of which the principal carotenoid is capsanthin. It has fair pH and heat stability, and poor light and chemical stability. It is used in sausages, meat products, condiment mixtures, and salad dressings.

Olive Oil   The oil obtained from the fruit of olive trees, *Olea europaea*. It is used mainly for salad and cooking oils.

Onion   A flavorant, the vegetable *Allium cepa* L., commercially processed into powder, salt, minced, and toasted forms. It is used in meats, sauces, soups, and dips.

Orange Terpenes   A density adjuster used to balance the oil phase in beverage emulsions.

Oregano   A spice made from the dried leaves of *Lippia graveolens*, a perennial of the mint family. There are two strains available. One strain, common to the Mediterranean region, is delicate in fragrance and taste and the other, which is common to Mexico, is quite pungent. It is used in sauces, soups, and pizza.

Orthophosphoric Acid   See **Phosphoric Acid**.

Oxidized Corn Starch   Starch produced by treating an aqueous starch suspension with dilute sodium hypochlorite containing a small excess of caustic soda (NaOH) until the desired degree of oxidation is reached. The slurry is then treated with an antichlor, such as sodium bisulfate, adjusted to the desired pH, filtered, washed, and dried. It still retains its original granule structure and is insoluble in water. It is extremely white, has decreased viscosity, is relatively clear, and shows a reduced tendency to thicken when cooled. Its food applications are those where high solids and low viscosity are desired.

Oxystearin   A crystallization inhibitor and release agent that is a modified fatty acid composed of the glycerides of partially oxidized stearic and other fatty acids. It is used in vegetable oils to prevent them from clouding in the refrigerator and in griddling fats and oils to prevent food from sticking to the frying pan.

P

Palmitic Acid $C_{16}H_{32}O_2$. A fatty acid which is a mixture of solid organic acids from fats consisting principally of palmitic acid with varying amounts of stearic acid. It functions as a lubricant, binder, and defoaming agent.

Palm Kernel Oil An oil obtained from palm kernels. It consists mainly of lauric, myristic, and oleic fatty acids. It resembles coconut oil and is used interchangeably with coconut oil. It is a possible source of stearine, which is a substitute for cocoa butter. It is used in margarine and confectionary.

Palm Oil The oil obtained from the fruit of the palm tree. It has a narrower plastic range than lard and most shortenings which is a disadvantage in shortening applications. It can be used in mixtures with only a moderately adverse effect on the plastic range. It consists mainly of palmitic, oleic, and linoleic fatty acids. It is used in margarine and shortenings.

Pantothenic Acid Vitamin B_5, which is a water-soluble vitamin. It is required for proper growth and maintenance of the body, and is involved in body processes such as energy release from carbohydrates and metabolism of fatty acids. It is relatively stable through storage, and is found in liver, eggs, and meat.

Papain A tenderizer that is a protein-digesting enzyme obtained from the papaya fruit. The enzyme, used in a patented process, is injected into the circulatory system of the live animal and is activated by the heat of cooking to break down the protein, thus tenderizing the beef. The enzyme is inactivated by stomach acids.

Paprika　A spice and colorant made from the ground, dried, ripe fruit of the herb *Capsicum annuum* L. It contributes flavor and color to foods. The pod provides red color and has good tinctorial strength, good pH stability, and poor stability to light and oxidation. It is used in meat, fish, sauces, and salad dressings. It is also termed sweet pepper or pimiento.

Parabens　Antimicrobial agents that are esters of para-hydroxybenzoic acid. The most common esters are methyl p-hydroxybenzoate and propyl p-hydroxybenzoate. The ethyl and butyl esters have some applications. It is related to benzoic acid and sodium benzoate but is effective over a wide pH range. The parabens are most active against yeasts and molds and are stable to high temperature. They are a white free-flowing powder of fair water solubility at room temperature which improves if the water is heated to 70°C. Methyl paraben is more soluble (0.25 g per 100 ml of water at 25°C) but less effective in mold inhibition than propyl paraben (0.04 g per 100 ml of water at 25°C). It is used in baked goods, beverages, and food color.

Para-Hydroxybenzoic Acid　See **Parabens**.

Parboiled Rice　The rice that results from the process of soaking rice in water, draining, pressure cooking to completely gelatinize the starch, drying, and milling. The parboiling process aids the development of stability toward cooking and heat processing. It is used in canned rice products such as soups, casseroles, meat, and rice dinners, such as Spanish rice. The milling of parboiled rice produces parboiled bran.

Parsley　A spice made from the dried leaves of *Petroselinum hortense*, of bright green color. It has a high content of vitamins A and C and also contain iron, iodine, copper, and manganese. It is used for garnishing and seasoning, with application in sauces, salads, and soups.

Partially Hydrogenated Coconut Oil　See **Coconut Oil**.

Partially Hydrogenated Oil　See **Hydrogenated Vegetable Oil**.

Pastry Flour　A flour obtained from soft wheat, either straight or clear flour grades because color is not an essential requirement. It is used in white sauces and pastry.

Patent Flour   Flour made from the separation of 40 to 90 percent of that portion of the grain that can be milled from a wheat blend. There are various streams to include long patent, medium patent, short patent, first patent, and fancy patent flours. See **Flour category**.

Peanut Oil   The oil obtained from peanuts, consisting principally of the unsaturated fatty acids oleic and linoleic. It is liquid at room temperature, has a specific gravity at 38°C of approximately 1.89 to 0.90, and an iodine number of 85 to 95. It is removed from the nuts by one of two processes, namely, the expeller method, in which the shelled peanuts are cooked with steam, and fed into an expeller press which physically presses the oil from the meal; or the pre-press solvent system, which is comparable to the expeller method except that less pressure is applied, which leaves more oil in the meal and the remaining meal is solvent-washed, usually with hexane, to dissolve the oil from the meal. The obtained crude oil is refined. The major use of peanut oil is in cooking oils and salad oils. Peanut oil is used in deep-fat frying because of its long frying life and high smoke point. In salad oil, it contributes to the suspension of solids. Other applications include shortening ingredient for doughnuts and cakes.

Pearl Starch   See **Cornstarch**.

Pectic Acid   Those pectic substances that are essentially void of methoxyl groups and have carboxyl groups only. They have varying degrees of neutralization. The divalent salts are slightly soluble in water and must be converted to the sodium or potassium forms for dissolution. It gels in the presence of calcium or other divalent cations.

Pectin   A gum that is water-soluble pectinic acid of varying methyl ester content and degree of neutralization. It is obtained from citrus peel and apple pomace. It forms a gel in systems of low pH (pH 2.8–3.7) and high sugar (55–80 percent) levels. The gel sets at 55 to 99°C and melts above 70°C. Pectins are characterized by rapid and slow set types. The high-methoxyl pectins have a degree of methylation (DM) greater than 50 percent, while those of less than 50 percent degree of methylation are termed low-methoxyl pectins. The low-methoxyl pectins gel in the presence of calcium ions and do not require a certain level of acid or sugar. It is used in beverages at 0.1 to 0.2 percent, in jams and jellies at 0.1 to 0.4 percent, and in confectioner's jelly at 0.8 to 1.5 percent.

Pectinic Acid A broad group of pectic substances that contain more than a negligible proportion of methyl ester groups and all the unesterified carboxyl groups are free. The divalent salts of pectinic acid are only slightly soluble in water and must be converted to the sodium or potassium form for dissolution. See **Pectin**.

Pentapotassium Triphosphate See **Potassium Tripolyphosphate**.

Pentasodium Tripolyphosphate See **Sodium Tripolyphosphate**.

Pepper A spice made from a berry from the vine *Piper nigrum* L. which produces black and white pepper. Black pepper is picked slightly underripe and dried, during which time the characteristic black, wrinkled appearance is attained. White pepper is picked fully ripe and dried, after which the outer hull is removed by attrition to expose the white core. It is used in meat, vegetables, soups, and salads.

Pepper, Cayenne A spice that is not related to the true pepper vine but to the paprika, bell peppers of the Capsicum family. It is hot and fiery, and used in spreads, dips, and chili sauce.

Pepper, Red The pod of the genus *Capsicum*, variety *C. annuum* L. and *C. frutescens* L. It has a hot, pungent flavor. It is used in barbecue sauce, spicy sauces, and chili powder.

Peptone A polypeptide used as a beer stabilizer.

Petrolatum A release agent, lubricant, and defoaming agent that is a purified mixture of semisolid hydrocarbons obtained from petroleum. It varies in color from white to yellow. It is used in bakery products, dehydrated fruits and vegetables, and egg white solids.

Petroleum Wax A wax used as a masticatory substance in chewing gum base. It is also used as a protective coating on raw fruits and vegetables and as a fruit defoamer.

Phenylethyl Anthranilate A synthetic flavoring agent that is a stable, white to yellow crystal of grape and orange blossom odor. It should be stored in glass or polyethylene-lined containers. It is used for flavors such as grape and cherry in applications such as beverages, ice cream, candy, and baked goods at 2 to 6 parts per million.

Phenylethyl Isobutyrate $C_{12}H_{16}O_2$. A synthetic flavoring agent that is a stable, colorless to light yellow liquid of fruity odor of floral note. It is soluble in alcohol and practically insoluble in water. It should be stored in glass or tin-lined containers. It is used in flavors for peach with applications in beverages, ice cream, candy, and baked goods at 3 to 13 parts per million.

Phosphate   Any salt of phosphoric acid. The salts include disodium phosphate, trisodium phosphate, sodium hexametaphosphate, and others. They play a variety of roles such as sequestrants, emulsifiers, solubility enhancers, and buffers in a variety of foods.

Phosphated Flour   Flour to which monocalcium phosphate is added at not less than 0.25 percent and not more than 0.75 percent. It is used in baked goods.

Phosphoric Acid H_3PO_4. An acidulant that is an inorganic acid produced by burning phosphorus in an excess of air, producing phosphorus pentoxide which is dissolved in water to form orthophosphoric acid of varying concentrations. It is a strong acid which is soluble in water. The acid salts are termed phosphates. It is used as a flavoring acid in cola and root beer beverages to provide desirable acidity and sourness. It is used as a synergistic antioxidant in vegetable shortenings. In yeast manufacture, it is used to maintain the acidic pH and provide a source for phosphorus. It also functions as an acidulant in cheese. It is also termed orthophosphoric acid.

Pimaricin   See **Natamycin**.

Pimiento   See **Paprika**.

Piperonyl Acetate   A synthetic flavoring agent that is a stable, colorless to light yellow liquid of heliotrope odor. It should be stored in glass or resin-lined containers. It is used in flavors for berry notes with applications in beverages, candy, ice cream, and baked goods at 50 to 90 parts per million.

Piperonyl Isobutyrate   A synthetic flavoring agent that is a moderately stable, colorless to light yellow liquid of fruity odor. It should be stored in glass or resin-lined containers. It is used in flavors for cherry, berry, and peach aroma with applications in beverages, candy, and baked goods at 1 to 4 parts per million.

Plasticizer   See **Softener**.

Polydextrose A bulking agent that is a randomly bonded condensation polymer of dextrose containing small amounts of bound sorbitol and citric acid. It is a water-soluble powder providing a pH range of 2.5 to 3.5. It is partially metabolized which results in a caloric value of one calorie per gram. As a reduced-calorie bulking agent, it can partially replace sugars and in some cases fats in reduced-calorie foods. It also functions as a bodying agent and humectant. Applications include desserts, specific baked goods, frozen dairy desserts, chewing gum, and candy. Usage levels vary according to application, but examples are frozen dessert, 13 to 14 percent; puddings, 8 to 9 percent; and cake, 15 to 16 percent.

Polyethylene A synthetic masticatory substance used in chewing gum.

Polyethylene Glycol Polyethylene glycols are addition polymers of ethylene oxide and water, ranging in molecular weight from 200 to 9500 and existing as liquids or solids. Those polyethylene glycols with a molecular weight below 600 are liquids which are miscible with water. Those polyethylene glycols with a molecular weight above 1000 are solids which are readily soluble in water. They function as flavoring adjuvants, lubricants, and dispersing and coating agents. They are used in carbonated beverages.

Polyglycerate 60 See **Ethoxylated Mono- and Diglycerides**.

Polyglycerol Esters of Fatty Acids Emulsifiers that are mixed partial esters formed by reacting polymerized glycerols with edible fats, oils, or fatty acids. They vary in degree of polymerization, and by varying the proportions and fats to be reacted, a diverse class of products is obtainable. The esters range from hydrophilic to lipophilic. They are used in cake mixes for volume and texture, in confectionary for gloss, in whipped toppings for aeration, and in flavors and colors as a solubilizer. Typical usage range is from 0.1 to 1.0 percent.

Polyoxyethylene (20) Mono- and Diglycerides of Fatty Acids See **Ethoxylated Mono- and Diglycerides**.

Polyoxyethylene Sorbitan Ester See **Polyoxyethylene Sorbitan Fatty Acid Esters**.

Polyoxyethylene Sorbitan Fatty Acid Esters Emulsifiers made by reacting ethylene oxide with sorbitan esters to increase their hydrophilic properties. They are generally used in oil and water emulsions in combination with lipophilic emulsifiers such as mono- and diglycerides or sorbitan monostearates to produce a wide variety of effects. They are also termed polysorbates, which include polysorbate 80 (polyoxyethylene [20] sorbitan monooleate), polysorbate 60 (polyoxyethylene [20] sorbitan monostearate), and polysorbate 65 (polyoxyethylene [20] sorbitan tristearate). They can solubilize essential and vitamin oils. They are used in panned coatings to reduce panning time, in coffee whiteners to prevent oiling-off, and in ice cream to produce dryness and overrun. Typical usage level ranges from 0.05 to 0.10 percent.

Polyoxyethylene (20) Sorbitan Monooleate An emulsifier produced by reacting oleic acid with sorbitol to yield a product which is reacted with ethylene oxide. It is a nonionic, water-dispersible surface-active agent that is very soluble in water. It is also termed polysorbate 80. It is used in ice cream and frozen desserts for overrun and dryness; as a disperser of flavors and colors in pickles; and for volume and texture in baked goods. It is frequently used with mono- and diglycerides at usage levels ranging from 0.05 to 0.10 percent.

Polyoxyethylene (20) Sorbitan Monostearate An emulsifier manufactured by reacting stearic acid with sorbitol to yield a product which is reacted with ethylene oxide. It is a nonionic, water-dispersible surface-active agent which is very hydrophilic. It is also termed polysorbate 60. It is used in whipped vegetable toppings for overrun and lightness; in cakes for increased volume and fine grain; in icings and confectionary for lightness and syneresis control; in salad dressing for emulsion stability. It is frequently used with sorbitan monostearate or mono- and diglycerides. The typical usage range is 0.10 to 0.40 percent.

Polyoxyethylene (20) Sorbitan Tristearate An emulsifier manufactured by reacting stearic acid with sorbitol to yield a product which is then reacted with ethylene oxide. It is a nonionic surface-active agent which is dispersible in fat, oil, and water. It is also termed polysorbate 65. It is used in frozen desserts, cakes, and coffee whiteners. It is frequently used with sorbitan monostearates or mono- and diglycerides. Typical usage range is 0.10 to 0.40 percent.

Polyoxyl (40) Stearate An emulsifier and antifoaming agent used in processed foods, fruit jellies, and sauces.

Polysorbates See **Polyoxyethylene Sorbitan Fatty Acid Esters**.

Polysorbate 60 See **Polyoxyethylene (20) Sorbitan Mono-stearate**.

Polysorbate 65 See **Polyoxyethylene (20) Sorbitan Tristearate**.

Polysorbate 80 See **Polyoxyethylene (20) Sorbitan Monooleate**.

Pomace Ground apple or fleshy fruit in the dry form.

Ponceau SX See **FD&C Red #4**.

Popcorn Indian corn that explodes when exposed to dry heat due to the expansion of the kernel.

Poppy Seed A spice that is a seed of *Papaver somniferum* L. Poppy seeds have a nutty flavor. They are used in breads, cakes, and butter sauce for vegetables, lending a nutlike flavor.

Potassium Acid Tartrate See **Cream of Tartar**.

Potassium Alginate $(C_6H_7O_6K)_n$. A gum that is the potassium salt of alginic acid. It is soluble in cold water, forming a viscous colloidal solution. It functions as a stabilizer, thickener, and gelling agent. It is used in dietetic foods, low-sodium foods, dry mixes, and dental impression material. Typical usage levels range from 0.05 to 0.50 percent.

Potassium Bicarbonate $KHCO_3$. An alkali and leavening agent obtained as colorless prisms or white powder. It is very soluble, with 1 g dissolving in 2.8 ml of water. Upon heating it liberates carbon dioxide which provides leavening in baked goods. It is also used in confectionary products.

Potassium Bisulfite A preservative that retards bacterial action, prevents discoloration, and functions as an antioxidant. It is not used in meats or in food recognized as a source of vitamin B_1, and it is not used on fruits or vegetables intended to be served raw or presented as fresh.

Potassium Bitartrate See **Cream of Tartar**.

Potassium Bromate $KBrO_3$. A dough conditioner that exists as white crystals or powder and is soluble in water. It exists in the anhydrous form as white granular powder and in the hydrated form as small white crystals or granules. It is used to age and improve the baking properties of flour. It is used with potassium iodate and azodicarbonamide to modify the protein in bread flour to promote the desired properties of loaf volume and shape. It is used in baked goods.

Potassium Carbonate K_2CO_3. A general purpose food additive and alkali. It is hygroscopic and the aqueous solutions are strongly alkaline. It has a solubility of 1 g in 1 ml of water at 25°C. It is used as a flavoring agent and processing aid, and to control pH. It is used in soups to neutralize acidity.

Potassium Carrageenan   See **Carrageenan**.

Potassium Caseinate   See **Caseinate**.

Potassium Chloride   KCl. A nutrient, dietary supplement, and gelling agent that exists as crystals or powder. It has a solubility of 1 g in 2.8 ml of water at 25°C and 1 g in 1.8 ml boiling water. Hydrochloric acid, and sodium chloride and magnesium chloride diminish its solubility in water. It is used as a salt substitute and mineral supplement. It has optional use in artificially sweetened jelly and preserves. It is used as a potassium source for certain types of carrageenan gels. It is used to replace sodium chloride in low-sodium foods.

Potassium Citrate, Monohydrate $C_6H_5K_3O_7 \cdot H_2O$. A sequestrant and buffer that exists as crystals or powder. It is slightly hygroscopic and possesses the advantageous properties of citric acid without having its acid reaction. A 1 percent solution has a pH of 7.5 to 9.0. It reacts with metal ions such as calcium, magnesium, and iron to form a complex. It is soluble in water with a solubility of 1.8 g in 20°C H_2O and 2 g in 1 ml of 80°C water. It is found in artificially sweetened jelly and in certain milk and meat products. Uses include processed cheese, puddings, and dietetic foods in which sodium is undesirable. It is also termed tripotassium citrate.

Potassium Dihydrogen Orthophosphate   See **Monopotassium Phosphate**.

Potassium Hydrogen Tartrate $KHC_4H_4O_6$. The acid potassium salt of tartaric acid, which has relatively poor water solubility. It functions to complex with heavy metal ions and regulates pH. See **Cream of Tartar**.

Potassium Hydroxide KOH. A water-soluble food additive and bleaching agent. Upon exposure to air it readily absorbs carbon dioxide and moisture and deliquesces. It is used to destroy the bitter chemical constituents in olives that will be used as black olives.

Potassium Iodate KIO_3. A source of iodine made by reacting iodine with potassium hydroxide. It is a crystalline powder which is more stable than iodide. It has a solubility of 1 g in 15 ml of water. It is used as a fast-acting dough improver; it is used with potassium bromate as an oxidizing agent to modify the protein in bread flour which promotes loaf volume and shape. It is used in baked goods.

Potassium Iodide KI. A source of iodine and a nutrient and dietary supplement. It exists as crystals or powder and has a solubility of 1 g in 0.7 ml of water at 25°C. It is included in table salt for the prevention of goiter.

Potassium Metabisulfite $K_2S_2O_5$. A chemical preservative and antioxidant obtained as white or colorless crystals, powder, or granules. It is soluble in water and insoluble in alcohol. The sulfite salt yields sulfurous acid at a low pH. It is used as a food preservative.

Potassium Metaphosphate $(KPO_3)_n$. A substitute for sodium phosphate in low-sodium foods. Also functions as a fermentation nutrient and buffer.

Potassium Nitrate KNO_3. A preservative and color fixative in meats which exists as colorless prisms or white granules or powder. It has a solubility of 1 g in 3 ml of water at 25°C. See **Nitrate**.

Potassium Nitrite KNO_2. A color fixative in meats which exists as white or yellowish granules or cylindrical sticks. It is very soluble in water. See **Nitrite**.

Potassium Oleate The potassium salt of oleic acid. It is used as a binder, emulsifier, and anticaking agent.

Potassium Orthophosphate A product used in chewing gum.

Potassium Palmitate The potassium salt of palmitic acid. It is used as a binder, emulsifier, and anticaking agent.

Potassium Phosphate Dibasic See **Dipotassium Phosphate**.

Potassium Phosphate Monobasic See **Monopotassium Phosphate**.

Potassium Phosphate Tribasic   See **Tripotassium Phosphate**.

Potassium Pyrophosphate   See **Tetrapotassium Pyrophosphate**.

Potassium Sodium Tartrate   See **Sodium Potassium Tartrate**.

Potassium Sorbate $C_6H_7KO_2$. A preservative that is the potassium salt of sorbic acid. It is a white crystalline powder which is very soluble in water, with a solubility of 139 g in 100 ml at 20°C. This solubility allows for solutions of high concentration which can be used for dipping and spraying. It is effective up to pH 6.5. It has approximately 74 percent of the activity of sorbic acid, therefore requiring higher concentrations to obtain comparable results as sorbic acid. It is effective against yeasts and molds and is used in cheese, bread, beverages, margarine, and dry sausage. Typical usage levels are 0.025 to 0.10 percent.

Potassium Stearate   The potassium salt of stearic acid. It is used as a binder, emulsifier, anticaking agent, and as a placticizer in chewing gum base.

Potassium Sulfite K_2SO_3. A preservative and antioxidant sulfite salt which yields free sulfurous acid at low pH. It is a white granular powder with a solubility of 1 g in 3.5 ml of water. It is used in a variety of food products.

Potassium Tripolyphosphate $K_5P_3O_{10}$. A phosphate that is very hygroscopic and very soluble in water with a solubility of 193 g in 100 ml at 25°C. It exists as white granules or powder and is mildly alkaline, with a pH of 10. It is used as a moisture binder in meat, as an emulsifier and suspending agent in food, and as a sequestrant for calcium and magnesium. It is also used in low-sodium products and chewing gum.

Potato Starch   A starch obtained from potatoes. It provides long body and clarity to food. It is used mainly in those countries in which it is the principal commercial starch. Applications include Danish desserts, soups, and gravies.

Powdered Sugar   A sweetener obtained by pulverizing granulated sugar and adding approximately 3 percent cornstarch. The blend is ground to the desired fineness, that is, 4X, 6X, or 8X. It is very soluble in water. Applications include confectionaries and icings.

Precipitated Calcium Phosphate   See **Tricalcium Phosphate**.

Pregelatinized Starch   Starch that has been processed to permit swelling in cold water, unlike natural starch which requires heating. The processing usually consists of cooking starch slurries, drying, and grinding to a fine powder. It is used in instant puddings, cake mixes, and soup mixes at 1 to 5 percent. It is also termed gelatinized wheat starch.

Propane   An aerating agent used in combination with chloropentafluoroethane or octafluorocyclobutane as a propellant and aerating agent for foamed or sprayed foods.

Propionic Acid $C_3H_6O_2$. The acid source of the propionates. Propionic acid in the liquid form has a strong odor and is corrosive, so it is used as the sodium, calcium, and potassium salts as a preservative. These yield the free acid in the pH range of the food in which they are used. It functions principally against mold. See **Calcium and Sodium Propionates**.

Propylene Glycol $C_2H_8O_2$. A humectant and flavor solvent that is a polyhydric alcohol (polyol). It is a clear, viscous liquid with complete solubility in water at 20°C and good oil solvency. It functions as a humectant, as do glycerol and sorbitol, in maintaining the desired moisture content and texture in foods such as shredded coconut and icings. It functions as a solvent for flavors and colors that are insoluble in water. It is also used in beverages and candy.

Propylene Glycol Alginate $(C_9H_{14}O_7)_n$. A gum that is the propylene glycol ester of alginic acid, which is obtained from kelp. As compared to sodium alginates, it has reduced sensitivity to acid and calcium salts. It functions in acidic systems. It functions as a thickener, stabilizer, and emulsifier in beer, salad dressings, syrups, and fruit drinks.

Propylene Glycol Dibenzoate   A clouding agent used for beverage emulsions.

Propylene Glycol Ester   See **Propylene Glycol Mono- and Di-Esters**.

Propylene Glycol Mono- and Di-Esters A lipophilic emulsifier that consists of propylene glycol esters of fatty acids, such as palmitic and stearic. It is used to increase the whipping ability and aeration in cake batters and whipped toppings. The usage level is the amount required to obtain the intended effect.

Propylene Glycol Monostearate A lipophilic emulsifier that is a propylene glycol ester. It is used as a dispersing aid in nondairy creamers; as a crystal stabilizer in cake shortenings and whipped toppings; and as an aeration increaser in cake batters, icings, and toppings. It is also used in oils and shortenings.

Propyl 2-Furanacrylate A synthetic flavoring agent that is a stable, colorless to light yellow liquid of fruity odor. It should be stored in glass or tin-lined containers. It is used in flavors for apple, pear, and raspberry with applications in beverages, candy, and baked goods at 1 to 3 parts per million.

Propyl Gallate $C_{10}H_{12}O_5$. An antioxidant used to retard fat and oil rancidity by imparting oxidative stability and thus protecting the flavor. It should be added to the fat or oil before oxidation has started. In direct addition, the propyl gallate is added slowly under vigorous agitation to the fat or oil which is heated to 60 to 70°C. It is often used with BHA and BHT because of the synergistic effect the three have in retarding rancidity. It is also used in combination with citric acid which acts as a metal chelating agent. It is susceptible to discoloration by certain metals such as iron or copper. It is used in fats, oils, meat products, potato sticks, and chewing gum.

Propyl Hepatanoate A synthetic flavoring agent that is a stable, colorless liquid of fruity odor. It should be stored in glass, tin, or resin-lined containers. It is used in apple flavors and modified coffee. It has applications in beverages, ice cream, candy, and baked goods at 4 to 18 parts per million.

Propyl p-Hydroxybenzoate See **Parabens**.

Propyl Paraben See **Parabens**.

Psyllium A gum obtained from the plant of the *Plantago* genus. It hydrates slowly to form a viscous dispersion of concentrations up to 1 percent. A clear, gelatinous mass is formed at 2 percent. It is used in bulk laxatives.

Pyridoxine   Vitamin B$_6$, a water-soluble vitamin with a solubility of 1 g in 5 ml of water. It functions in the utilization of protein and for proper growth, and is an essential nutrient in enzyme reactions. During processing, there is a loss due to leaching of the vitamin in water. It is destroyed by high temperatures, high irradiation, and exposure to light. During storage, loss increases with temperature and storage time. It is found in liver, eggs, and meats. It is also termed pyridoxine hydrochloride.

Pyridoxine Hydrochloride   C$_{18}$H$_{11}$NO$_3$·HCl. Vitamin B$_6$, a water-soluble vitamin. It is soluble in water, and slightly soluble in alcohol. It is slowly affected by sunlight and is reasonably stable in air. It has a pH of 2.3 to 3.5. It is also termed vitamin B$_6$ hydrochloride. See **Pyridoxine**.

Quicklime   See **Calcium Oxide**.

Quillaia   A foaming agent extracted from the soap bark tree *Quillaia saponaria*. It has no discernible color, odor, flavor, or aftertaste. It is used as a foaming agent in root beer and other beverages.

Quince Seed   A gum produced from the fruit of the quince tree *Cydonia oblonga*. It hydrates slowly to form a highly viscous dispersion at concentrations up to 1.5 percent. Above 2 percent, a slimy, muscilaginous mass is formed. It is principally used in the cosmetic industry. It is also termed gum quince seed, semen cydonia, golden apple seed, and cydonia seed.

Quinine $C_{20}H_{24}N_2O_2$. A flavorant naturally obtained from the cinchona tree. It is used as a bitter flavoring in beverages such as quinine water, tonic water, and bitter lemon. Quinine sulfate and quinine hydrochloride are cleared for use as a flavor in carbonated beverages at levels less than 83 parts per million.

R

Raisin   A dried grape used as a fruit and as an ingredient in cereals, baked goods, and desserts.

Raisin Seed Oil   See **Grape Seed Oil**.

Rape Seed Oil   The oil derived from seeds of *Brassica campestris* and related trees. It can function as a stabilizer and thickener in peanut butter and as an emulsifier in cake mix shortenings. It is principally used in the United States as an additive for lubricants.

Raw Sugar   A sweetener that is an intermediate product, containing nonsugar impurities, thus being less refined than white sugar. It is made by crushing and shredding sugar cane to extract the juice which is processed to yield raw sugar and upon further processing yields refined cane sugar.

Red Durum Wheat   Wheat obtained from the durum wheat kernel. It is used in macaroni and spaghetti products. See **Durum Wheat**.

Reducing Sugar   A sugar that can chemically react with copper in an alkaline solution. It combines with nitrogen compounds at elevated temperature to produce a browning "Maillard" reaction which contributes to the production of a brown crust in baked goods. It is used in the production of caramel color. Dextrose and fructose are reducing sugars.

Rennet   A milk coagulant that is the concentrated extract of rennin enzyme obtained from calves' stomachs (calf rennet) or adult bovine stomachs (bovine rennet). The commercial saline extract of rennin contains a little pepsin, some sodium chloride, and some boric acid, sodium benzoate, or propylene glycol as a preservative. In the paste form, it also contains lipase. In the paste form it is used in Italian-type cheeses. It is used to coagulate milk into curd in making cheese and junket. A microbial rennet and a pepsin rennet also exist. See **Rennin**.

Rennet Casein   The product that results from the precipitation of pasteurized milk with a rennet enzyme. Rennet casein requires a pH above 9 to dissolve, as compared to acid casein, which can be dissolved in alkali at a pH as low as 6.5. Rennet casein can be dispersed at lower pH by adding a complex phosphate such as sodium tripolyphosphate. This results in a casein of good emulsifying, whipping, foam stability, and water-binding properties. Uses include imitation cheese.

Rennin   A milk coagulant that is an enzyme obtained from the abomasum portion of the stomach of suckling mammals. It is most active at pH 3.8. One part purified rennin will coagulate more than five million parts of milk. The commercial extract of rennin is termed rennet. It is used to coagulate milk in making cheese, junket, and custard. See **Rennet**.

Retinol $C_{20}H_{30}O$. The fat-soluble vitamin A which is required for new cell growth and prevention of night blindness. There is no appreciable loss by heating or freezing and it is stable in the absence of air. Sources include liver, fortified margarine, egg, and milk. Vitamin A palmitate can be found in frozen egg substitute.

Riboflavin $C_{17}H_{20}N_4O_6$. The water-soluble vitamin B_2, required for healthy skin and the building and maintaining of body tissues. It is a yellow to orange-yellow crystalline powder. It acts as a coenzyme and carrier of hydrogen. It is stable to heat but may dissolve and be lost in cooking water. It is relatively stable to storage. Sources include leafy vegetables, cheese, eggs, and milk.

Rice Bran Oil   An oil made from rice bran that consists mainly of oleic, linoleic, and palmitic fatty acids. It is used in salad oil, cooking oil, and hydrogenated shortenings.

Rice Bran Wax   A refined wax obtained from rice bran. It is insoluble in water. It is used in candy, fresh fruits, and vegetables as a coating and as a plasticizing material in chewing gum.

Rice Flour   The flour made from different varieties of long-, medium-, and short-grain rice, usually obtained from the broken milled rice. The chemical composition is the same as that of the whole rice. The flour does not contain gluten and, as a result, doughs made from it do not retain the gases generated during baking. Rice flours from different varieties display characteristic viscosity patterns during the heating and cooling of their pastes. In general, rice with starch of an amylose content greater than 22 percent has a relatively low peak viscosity and forms a rigid gel on cooling (high set-back viscosity). Rice with a starch low in amylose has a high peak and low set-back viscosity. Rice flour is used in formulated baby foods, breakfast foods, meat products, and breading.

Rice Starch   The starch obtained from rice. It forms tender, opaque gels. It has some use in puddings.

Rochelle Salt   See **Sodium Potassium Tartrate**.

Roselle   A natural red colorant obtained from the roselle flower extract from hibiscus. It does not contain tannins as do other anthocyanins such as grape skin extract. It has good stability at pH 3, and changes from red to brown to green as the pH approaches neutral. It has fair heat, light, and chemical stability with low tinctorial strength. It is permitted only in beverages.

Rosemary   A spice made from the dried leaves of *Rosmarinus officinalis* L., an evergreen shrub. It has a medicinal, menthol flavor. It is available in whole and ground forms. It is used in soups, poultry, and meats, especially lamb.

Rum Ether   A synthetic flavoring agent that is a stable, colorless to yellow liquid of ethereal rum-like note. It should be stored in glass and stainless steel containers. It is used to intensify rum flavors for application in beverages, candy, and ice cream at 67 to 320 parts per million and in alcoholic beverages at 1600 parts per million. It is also termed ethyl oxyhydrate.

Rye   A cereal crop that is a source of rye flour. It is used as a bread grain.

Rye Flour   The flour obtained by milling rye. It is available in white, medium, and dark grades and has a distinct flavor. It is usually diluted with wheat flour in order to make it more palatable. It is used in bread making.

Saccharin $C_7H_5NO_3S$. A non-nutritive synthetic sweetener which is 300 to 400 times sweeter than sucrose. It is nonhygroscopic and has a bitter aftertaste and a stability problem in cooked, canned, or baked goods. It is slightly soluble in water with a solubility of 10 g in 100 g of water at 25°C, but the solubility improves in boiling water. As sodium saccharin, there are two forms, $C_7H_4NNaO_3S \cdot 2H_2O$ (chemical name: 1,2-benzisothiazolin-3-one-1, 1-dioxide, sodium salt dihydrate), with a solubility of 1 g in 1.2 ml H_2O; and $C_7H_4NNaO_3S$ (chemical name: 1,2-benzisothiazolin-3-one-1,1-dioxide, sodium salt). Calcium saccharin $(C_7H_4NO_3S)_2Ca$ (chemical name: 1,2-benzisothiazolin-3-one-1,1-dioxide, calcium salt) is used where low sodium content and reduced aftertaste are required. It is used in low-calorie foods such as jam, beverages, and desserts. It is also termed sodium benzosulfimide.

Saccharose See **Sugar**.

Safflower Oil An unsaturated oil obtained from the safflower seed of the plant *Carthamus tinctorius*. It consists mainly of linoleic and oleic fatty acids. It is used principally as a drying oil in the United States.

Saffron A spice obtained from the dried stigmas of the fall-flowering *Crocus sativus* L. The flower stigma is of intense yellow color. It has a powerful, somewhat bitter aroma. It is used in breads, fish, chicken, sauces, and rice dishes.

Sage A spice made from the dried leaves of the shrub *Salvia officinalis*

L. It has a strong, fragrant odor. It is available industrially as whole leaf, cut, rubbed, and ground to determined granulations. It is used in pork, soups, poultry seasonings, and fish.

Sago Starch   The starch obtained from the sago palm *Metroxylon sagus* or *M. rumphii* and the palm fern *Cycas circinalis*. It forms high-strength gels which lose their clarity upon standing. It is used in confections and puddings.

Saint John's Bread   See **Locust Bean Gum**.

Salt   NaCl. A seasoning and preservative whose chemical composition is sodium chloride, about 40 percent sodium and 60 percent chlorine by weight. It contains not less than 97.5 percent sodium chloride after drying, while high-grade salt contains 99.8 percent sodium chloride. Salt production can be by solar evaporation, rock salt mining, and vacuum pan evaporation. The method selected depends on climate, character of the deposit, and type of salt required. Seasoned salt contains added flavors. It is available in several particle sizes (coarse, flake, fine), shapes (flake, cube) which relate to density, solubility, flow, blending, and adherance. It is used as a carrier for dry or semidry ingredients or as an ingredient in prepared mixes. It is used in cheese, butter, and salted nuts for flavor. It is used in cheese manufacture to help remove the whey and suppress the growth of unwanted organisms, in sausage as a seasoning and curing agent, and in baked goods, pickles, and sauerkraut for flavor and fermentation control.

SAPP   See **Sodium Acid Pyrophosphate**.

Savory   A spice that is the dried leaves and flowering tops of the plant *Satureia hortensis* L. The two distinct varieties are summer savory and winter savory. Summer savory is generally preferred because it has a more delicate flavor and is less resinous. It is used in soups, salads, and sauces.

Sea Salt   See **Salt**.

Seasoned Salt   See **Salt**.

Self-Rising Flour   White flour to which sodium bicarbonate and one or more of the acid-reacting substances are added, that is, monocalcium phosphate, sodium acid pyrophosphate, or sodium aluminum phosphate. It is seasoned with salt. The inclusion of these ingredients provides a leavening system that allows the flour to rise when wetted in the preparation of baked goods.

Semen Cydonia   See **Quince Seeds**.

Semolina   The purified ground middlings of durum wheat. It contains bran specks. Durum semolina is ground so that not more than 3 percent passes through a number 100 U.S. sieve. It takes longer to cook and is more resistant to overcooking than flour and results in less cloudiness in the water. It has a 50 percent relative protein efficiency as compared to nonfat dry milk. It is used in macaroni and spaghetti products.

Sesame Oil   The oil obtained from sesame seeds. It consists principally of oleic and linoleic fatty acids. It has resistance to oxidation. It is used in vegetable shortenings, salad oil, and cooking oil, and is found in frozen chicken chow mein.

Sesame Seed   The seed of the plant *Sesamum indicum* L. It has a sweet, "nutty" flavor. It yields sesame oil. It is used in breads, meats, and vegetables. It is also termed benne.

Shallot   A spice, *Allium ascalonicum*, a member of the onion family. It ranges in size from walnut to small fig and is milder than the onion. It can be substituted for the onion and is used in sauces, dressings, soups, and meats.

Shortening   Any animal or vegetable fat or oil that "shortens" or retards the development of gluten strands in baked goods for the purpose of producing a tender, crisp texture. Solid fats are most commonly used instead of oils because of their plastic nature. It is used in baked goods.

Silica, Amorphous   See **Silicon Dioxide**.

Silicon Dioxide SiO_2. An anticaking agent, carrier, and dispersant that can absorb approximately 120 percent of its weight and remain free flowing. It is used in salt, flours, and powdered soups to prevent caking caused by moisture. It is also used in powdered coffee whitener, vanilla powder, baking powder, dried egg yolk, and tortilla chips. The usage level ranges from 1 to 2 percent. It is also termed silica, amorphous.

Skeletal Meat   The edible part of the animal that is muscle tissue attached to the bone. It includes the shoulder and side of pork, brisket, flank, and round of beef. It is an ingredient in sausage.

Skim Milk   Milk from which sufficient fat has been removed to reduce the milkfat content to less than 0.5 percent. It is used in the manufacture of certain cheese varieties, casein, and lactose. It is an ingredient in frozen desserts, baked goods, and confectionary. It is also consumed as a beverage.

Skim Milk Powder   See **Milk Solids – Not-Fat**.

Slaked Lime $Ca(OH)_2$. See **Calcium Hydroxide**.

Smoke Flavoring   A flavorant that can be obtained in the form of liquid smoke derived from burning hardwoods such as maple and hickory or as synthetic smoke made by synthesis. It is used to impart flavor and aroma to bacon, ham, and sausage.

Soap Bark   See **Quillaia**.

Soda Alum   See **Aluminum Sodium Sulfate**.

Sodium   A metal element that performs bodily functions.

Sodium Acetate $C_2H_3NaO_2 \cdot 3H_2O$. A source of acetic acid that is obtained as crystals or powder. It has a solubility in water of 1 g in 0.8 ml of water.

Sodium Acetate, Anhydrous $C_2H_3NaO_2$. A source of acetic acid obtained as a granular powder. It has a solubility in water of 1 g in 2 ml of water.

Sodium Acid Carbonate   See **Sodium Bicarbonate**.

Sodium Acid Phosphate   See **Monosodium Phosphate**.

Sodium Acid Pyrophosphate $Na_2H_2P_2O_7$. A leavening agent, preservative, sequestrant, and buffer which is mildly acidic with a pH of 4.1. It is moderately soluble in water, with a solubility of 15 g in 100 ml at 25°C. It is used in doughnuts and biscuits for its variable gas release rate during the mixing, bench action, and baking process. It is used in baking powder as a leavening agent. It is used in canned fish products to reduce the level of undesired struvite crystals (magnesium ammonium phosphate hexahydrate) by complexing the magnesium. It is used to sequester metals in processed potatoes. It it also termed SAPP, disodium dihydrogen pyrophosphate, acid sodium pyrophosphate, disodium diphosphate, and sodium pyrophosphate.

Sodium Alginate $(C_6H_7O_6Na)_n$. A gum obtained as a sodium salt of alginic acid, which is obtained from seaweed. It is cold- and hot-water soluble, producing a range of viscosities. It forms irreversible gels with calcium salts or acids. It functions as a thickener, binder, and gelling agent in dessert gels, puddings, sauces, toppings, and edible films.

Sodium Aluminosilicate See **Sodium Silicoaluminate**.

Sodium Aluminum Phosphate, Acidic $NaH_{14}Al_3(PO_4)_8 \cdot 4H_2O$. A leavening agent, slowly soluble in water which gives it a delayed leavening reaction. It has a pH of 2.8. Approximately 20 percent of the carbon dioxide is released during the mixing period and the remainder is released during the baking period when the batter is exposed to heat. It has a high tolerance to variation in batter preparation. It is used in prepared mixes such as cake and pancake mixes.

Sodium Aluminum Phosphate, Basic $Na_8Al_2(OH)_2(PO_4)_4$. An emulsifier that is a white powder which is barely soluble in water. It has a pH of 9.2. It may be used in processed cheese to provide consistency and to aid in eliminating surface crystals.

Sodium Aluminum Sulfate A leavening agent that releases the majority of the gas during baking, and is not used alone but in combination with a faster-acting leavening agent such as monocalcium phosphate. This results in a double-acting baking powder. It is almost nonreactive until heat is applied. It is used in baked goods.

Sodium Ascorbate $C_6H_7NaO_6$. An antioxidant that is the sodium form of ascorbic acid. It is soluble in water and provides a non-acidic taste. A 10 percent aqueous solution has a pH of 7.3 to 7.6. In water, it readily reacts with atmospheric oxygen and other oxidizing agents, making it valuable as an antioxidant. One part sodium ascorbate is equivalent to 1.09 parts of sodium erythorbate. See **Ascorbic Acid**.

Sodium Benzoate $C_7H_5NaO_2$. A preservative that is the sodium salt of benzoic acid. It converts to benzoic acid, which is the active form. It has a solubility in water of 50 g in 100 ml at 25°C. Sodium benzoate is 180 times as soluble in water at 25°C as is the parent acid. The optimum functionality occurs between pH 2.5 to 4.0 and it is not recommended above pH 4.5. It is active against yeasts and bacteria. It is used in acidic foods such as fruit juices, jams, relishes, and beverages. Its use level ranges from 0.03 to 0.10 percent.

Sodium Benzosulfimide   See **Saccharin**.

Sodium Bicarbonate $NaHCO_3$. A leavening agent with a pH of approximately 8.5 in a 1 percent solution at 25°C. It functions with food grade phosphates (acidic leavening compounds) to release carbon dioxide which expands during the baking process to provide the baked good with increased volume and tender eating qualities. It is also used in dry-mix beverages to obtain carbonation, which results when water is added to the mix containing the sodium bicarbonate and an acid. It is also termed baking soda and sodium acid carbonate.

Sodium Biphosphate   See **Monosodium Phosphate**.

Sodium Bisulfate $NaHSO_4$. A strong acid that exists as white crystals or granules and is soluble in water.

Sodium Bisulfite $NaHSO_3$. A preservative that exists as a powder, with a solubility of 1 g in 4 ml of water. It prevents discoloration and inhibits bacterial growth. It is used in dried fruit to inhibit browning and maintain the bright color. It is found in reconstituted lemon juice. See **Sulfur Dioxide**.

Sodium Calcium Alginate   A gum that is the sodium and calcium salt of alginic acid. It functions as a thickener, gel former, and stabilizer in a variety of foods.

Sodium Calcium Aluminosilicate, Hydrated   An anticaking agent for use at levels not to exceed 2 percent. It is also termed sodium calcium silicoaluminate.

Sodium Calcium Silicoaluminate   See **Sodium Calcium Aluminosilicate, Hydrated**.

Sodium Caprate   The sodium salt of capric acid. It functions as a binder, emulsifier, and anticaking agent.

Sodium Caprylate   The sodium salt of caprylic acid. It functions as a binder, emulsifier, and anticaking agent.

Sodium Carbonate Na_2CO_3. An alkali that exists as crystals or crystalline powder and is readily soluble in water. It has numerous functions: as an antioxidant, as a curing and pickling agent, as a flavoring agent, a processing aid, a sequestrant, and as an agent for pH control. It is used in instant soups to neutralize acidity. It is used in alginate water dessert gels to sequester the calcium, allowing the alginate to solubilize. It is also used in puddings, sauces, and baked goods.

Sodium Carboxymethylcellulose   See **Carboxymethylcellulose**.

Sodium Carrageenan   See **Carrageenan**.

Sodium Caseinate   The sodium salt of casein, a milk protein. It is used as a protein source and for its functional properties such as water binding, emulsification, whitening, and whipping. It is used in coffee whiteners, nondairy whipped toppings, processed meat, and desserts.

Sodium Chloride   See **Salt**.

Sodium Citrate   A buffer and sequestrant obtained from citric acid as sodium citrate anhydrous ($C_6H_5Na_3O_7$) and as sodium citrate dihydrate or sodium citrate hydrous ($C_6H_5Na_3O_7 \cdot 2H_2O$). The crystalline products are prepared by direct crystallization from aqueous solutions. Sodium citrate anhydrous has a solubility in water of 57 g in 100 ml at 25°C, while sodium citrate dihydrate has a solubility of 65 g in 100 ml at 25°C. It is used as a buffer in carbonated beverages and to control pH in preserves. It improves the whipping properties in cream and prevents feathering of cream and nondairy coffee whiteners. It provides emulsification and solubilizes protein in processed cheese. It prevents precipitation of solids during storage in evaporated milk. In dry soups, it improves rehydration which reduces the cooking time. It functions as a sequestrant in puddings. It functions as a complexing agent for iron, calcium, magnesium, and aluminum. Typical usage levels range from 0.10 to 0.25 percent. It is also termed trisodium citrate.

Sodium Dehydroacetate $C_8H_7NaO_4 \cdot H_2O$. A preservative that is a white powder with a solubility of 1 g in 3 ml of water. It is the sodium salt of dehydroacetic acid.

Sodium Diacetate $C_4H_7NaO_4 \cdot xH_2O$. A preservative, sequestrant, acidulant, and flavoring agent that is a molecular compound of sodium acetate and acetic acid which yields acetic acid. It is a white crystalline powder which is hygroscopic. It functions against mold and bacteria and is used in bread.

Sodium Dioctylsulfosuccinate See **Dioctyl Sodium Sulfosuccinate**.

Sodium Erythorbate $C_6H_7NaO_6 \cdot H_2O$. An antioxidant that is the sodium salt of erythorbic acid. In the dry crystal state, it is nonreactive but in water solution it readily reacts with atmospheric oxygen and other oxidizing agents, a property that makes it valuable as an antioxidant. During preparation, a minimal amount of air should be incorporated and it should be stored at a cool temperature. It has a solubility of 15 g in 100 ml of water at 25°C. On a comparative basis, 1.09 parts of sodium erythorbate are equivalent to 1 part of sodium ascorbate; 1.23 parts of sodium erythorbate are equivalent to 1 part erythorbic acid. It functions to control oxidative color and flavor deterioration in a variety of foods. In meat curing, it controls and accelerates the nitrite curing reaction and maintains the color brightness. It is used in frankfurters, bologna, and cured meats and is occasionally used in beverages, baked goods, and potato salad.

Sodium Ferrocyanide See **Yellow Prussiate of Soda**.

Sodium Ferric Pyrophosphate See **Sodium Iron Pyrophosphate**.

Sodium Hexametaphosphate $(NaPO_3)_{13}$. A sequestrant and moisture binder that is very soluble in water but dissolves slowly. Solutions have a pH of 7.0. It permits peanuts to be salted in the shell by making it possible for the salt brine to penetrate the peanuts. In canned peas and lima beans, it functions as a tenderizer when added to the water used to soak or scald the vegetables prior to canning. It improves whipping properties in whipping proteins. It functions as a sequestrant for calcium and magnesium, having the best sequestering power of all the phosphates. It prevents gel formation in sterilized milk. It is also termed sodium metaphosphate, sodium polyphosphate, and Graham's salt.

Sodium Hydrogen Carbonate See **Sodium Bicarbonate**.

Sodium Hydrogen Diacetate See **Sodium Diacetate**.

Sodium Hydrogen Malate   An acidulant.

Sodium Hydroxide   NaOH. An alkali that is soluble in water, having a solubility of 1 g in 1 ml of water. It is used to destroy the bitter chemicals in olives that are to become black olives. It also functions to neutralize acids in various food products.

Sodium Hyposulfite   See **Sodium Thiosulfate**.

Sodium Iron EDTA   See **Iron**.

Sodium Iron Pyrophosphate $Na_8Fe_4(P_2O_7)_5 \cdot xH_2O$. A nutrient and dietary supplement that is a source of iron. It contains approximately 14.5 percent iron and is insoluble in water. It is utilized for the enrichment of foods that are susceptible to rancidity. It is also termed sodium ferric pyrophosphate.

Sodium Isoascorbate   See **Sodium Erythorbate**.

Sodium Lactate $C_3H_5NaO_3$. A humectant that is the sodium salt of lactic acid which is low-melting and hygroscopic with a mildly saline taste. It is used in sponge cake and Swiss roll to produce a tender crumb and to reduce staling. It provides a protein plasticizing effect in biscuits. It is used in frankfurter-type sausages as a replacement for sodium chloride and as a dehydrating salt or humectant in uncured hams.

Sodium Laurate   The sodium salt of lauric acid. It functions as a binder, emulsifier, and anticaking agent.

Sodium Lauryl Sulfate   An emulsifier and whipping aid that has a solubility of 1 g in 10 ml of water. It functions as an emulsifier in egg whites. It is used as a whipping aid in marshmallows and angel food cake mix. It also functions to aid in dissolving fumaric acid.

Sodium Metabisulfite $Na_2S_2O_5$. A preservative and antioxidant that exists as crystals or powder having a sulfur dioxide odor. It is readily soluble in water. It is used in dried fruits to preserve flavor, color, and to inhibit undesirable microorganism growth. It prevents "black spots" due to oxidative deterioration in shrimp. It is used in maraschino cherries. It is found in lemon drinks as a preservative. See **Sulfur Dioxide**.

Sodium Metaphosphate   See **Sodium Hexametaphosphate**.

Sodium Myristate   The sodium salt of myristic acid. It functions as a binder, emulsifier, and anticaking agent.

Sodium Nitrate $NaNO_3$. The salt of nitric acid that functions as an antimicrobial agent and preservative. It is a naturally occurring substance in spinach, beets, broccoli, and other vegetables. It consists of colorless, odorless crystals or crystalline granules. It is moderately deliquescent in moist air and is readily soluble in water. It is used in meat curing to develop and stabilize the pink color. See **Nitrate**.

Sodium Nitrite $NaNO_2$. The salt of nitrous acid that functions as an antimicrobial agent and preservative. It is a slightly yellow granular powder or nearly white, opaque mass or sticks. It is deliquescent in air. It has a solubility of 1 g in 1.5 ml of water. It is used in meat curing for color fixation and development of flavor. See **Nitrite**.

Sodium Oleate   The sodium salt of oleic acid. It functions as a binder, emulsifier, and anticaking agent.

Sodium Palmitate   The sodium salt of palmitic acid. It functions as a binder, emulsifier, and anticaking agent.

Sodium Pectinate   See **Pectin**.

Sodium Phosphate, Dibasic   See **Disodium Phosphate**.

Sodium Phosphate, Dibasic Dihydrate   See **Disodium Phosphate, Duohydrate**.

Sodium Phosphate, Monobasic   See **Monosodium Phosphate**.

Sodium Phosphate, Tribasic   See **Trisodium Phosphate**.

Sodium Polyphosphate $(NaPO_3)_n$. A sequestrant and emulsifier that has a variety of functions in foods. It is used in cheese, dairy products, meat, fish, and poultry. It is specifically termed sodium tetrametaphosphate, sodium hexametaphosphate, and Graham's salt as determined by the chain length.

Sodium Potassium Tartrate $C_4H_4KNaO_6 \cdot 4H_2O$. A buffer and sequestrant that is the salt of $L(+)$ − tartaric acid. It has a solubility in water of 1 g in 1 ml. It is also termed Rochelle salt.

Sodium Propionate $C_3H_5NaO_2$. A preservative that is the salt of propionic acid. It is very soluble in water with a solubility of 100 g in 100 ml of water at 25°C. It is most effective against mold, with little activity against bacteria, and no activity against yeasts. Its optimum effectiveness is up to pH 5.0. Sodium propionate is more soluble than calcium propionate. It is used in baked goods to prevent mold and rope. Usage level ranges from 0.1 to 0.4 percent based on the weight of the flour.

Sodium Pyrophosphate See **Sodium Acid Pyrophosphate**.

Sodium Pyrophosphate, Tetrabasic See **Tetrasodium Pyrophosphate**.

Sodium Saccharin See **Saccharin**.

Sodium Silicate Na_2SiO_3. A product used as a preservative for eggs.

Sodium Silicoaluminate An anticaking and conditioning agent used to improve flow properties and prevent caking. It absorbs moisture up to 75 percent of its weight. It functions as a moisture absorbent, moisture barrier, carrier, and processing aid. It is used in salt, cake mixes, powdered sugar, nondairy creamers, and dry mixes. Usage level ranges from 1 to 2 percent. It is also termed sodium aluminosilicate.

Sodium Sorbate A preservative that is the salt of sorbic acid. It is partially soluble in water and is used effectively against yeasts and molds up to pH 6.5. It is not usually used as a replacement for sorbic acid or potassium sorbate. It is used in cheese and baked goods.

Sodium Stearate The sodium salt of stearic acid. It functions as a binder, emulsifier, and anticaking agent. It is used as a plasticizer in chewing gum base.

Sodium Stearoyl Fumarate A conditioning agent that functions as a dough conditioner for yeast-raised baked goods. It is also used as a conditioning agent for dehydrated potatoes.

Sodium Stearoyl-2-Lactylate See **Stearoyl Lactylate**.

Sodium Stearyl Fumarate A dough conditioner and conditioning agent that is a white powder practically insoluble in water. It is used as a dough conditioner in yeast-raised baked goods. It is used as a conditioning agent in dehydrated potatoes. It also functions as a maturing and bleaching agent.

Sodium Sulfate Na_2SO_4. The salt of sulfuric acid that is readily soluble in water and exists as crystals or crystalline powder. It is used in caramel production.

Sodium Sulfite   See **Sulfur Dioxide.**

Sodium Tartrate $Na_2C_4H_4O_6 \cdot 2H_2O$. A sequestrant and stabilizer that is the disodium salt of L(+) − tartaric acid. It is soluble in water. It functions as a sequestrant and stabilizer in meat products and sausage casings. It is also termed disodium tartrate.

Sodium Tetrametaphosphate $(NaPO_3)_6$. A sequestrant and emulsifier that is infinitely soluble in water. It is used as a water binder in cured pork. It is also termed Graham's salt and sodium polyphosphate.

Sodium Thiosulfate $Na_2S_2O_3 \cdot 5H_2O$. A sequestrant, antioxidant, and formulation aid that is a powder soluble in water. It can be used in alcoholic beverages at 5 parts per million and in table salt at 0.1 percent. It is also termed sodium hyposulfite.

Sodium Triphosphate   See **Sodium Tripolyphosphate**.

Sodium Tripolyphosphate $Na_5P_3O_{10}$. A binder, stabilizer, and sequestrant that is mildly alkaline, with a pH of 10 and moderately soluble in water, with a solubility of 15 g in 100 ml of water at 25°C. It is used to improve the whipping properties of egg-containing angel food cake mix and meringues. It reduces gelling of juices and canned ham and tenderizes canned peas and lima beans. It is a moisture binder in cured pork and protects against discoloration and reduces shrinkage in sausage products. In algin desserts, it functions as a calcium sequestrant. It is also termed pentasodium tripolyphosphate and sodium triphosphate.

Softener   A term used for ingredients that soften. Softening relates to the hygroscopicity and the ability of the polyhydric alcohol, such as propylene glycol or glycerin, to retain moisture. Softeners are used in shredded coconut, pet foods, and chewing gum to maintain moistness. It is also termed plasticizer.

Sorbic Acid $C_6H_8O_2$. A preservative that is effective against yeasts and molds. It is effective over a broad pH range up to pH 6.5, being ineffective above pH 7.0. It is a white, free-flowing powder which is slightly soluble in water with a solubility of 0.16 g in 100 ml of water at 20°C. Its solubility in water increases with increasing temperatures, although it is not recommended in foods that are pasteurized because it breaks down at high temperatures. The salts are potassium, calcium, and sodium sorbate. It is used in cheese, jelly, beverages, syrup, and pickles. Typical usage levels range from 0.05 to 0.10 percent.

Sorbitan Ester   A lipophilic emulsifier whose permitted type in foods is sorbitan monostearate. It is used in cakes, chocolate, and coffee whitener. It is also termed sorbitan fatty acid ester.

Sorbitan Fatty Acid Ester   See **Sorbitan Ester**.

Sorbitan Monostearate   A lipophilic emulsifier that is a sorbitan fatty acid ester, being a sorbitol-derived analog of glycerol monostearate. It is a nonionic, oil-dispersible surface-active agent. It is used as a gloss enhancer in chocolate coatings; as a dispersant aid in coffee whiteners; to increase volume in cakes and icings; and often in combination with polysorbates. Typical usage level ranges from 0.30 to 0.70 percent.

Sorbitol $C_6H_{14}O_6$. A humectant that is a polyol (polyhydric alcohol) with good solubility in water and poor solubility in oil. It is approximately 60 percent as sweet as sugar. It maintains moistness in shredded coconut, pet foods, and candy. It is used in low-calorie beverages to provide body and taste. It is used in dietary foods such as sugarless candy and chewing gum. It is also used as a crystallization modifier in soft sugar-based confections.

Sorghum Oil   An oil consisting mainly of linoleic and oleic fatty acids. It is similar in composition and properties to corn oil.

Soy Flour   The powdered product obtained from defatted soybean. It has 50 percent or more protein content. It is used in doughnuts, bread, cereals, and sausage products as a nutrient and binder.

Soy Flour, Lecithinated   See **Lecithinated Soy Flour**.

Soy Flour, Textured   See **Textured Soy Flour**.

Soy Oil   See **Soybean Oil**.

Soy Protein See **Soybean Protein**.

Soy Protein Concentrate See **Soybean Protein Concentrate**.

Soy Protein Isolate See **Soybean Protein Isolate**.

Soybean A legume of high protein content, containing 40 percent or greater protein and approximately 18 percent oil. The protein contains all the essential amino acids. Soybeans are processed to produce soybean flour, protein concentrate, protein isolate, and soybean oil.

Soybean Flour The flour made from defatted soybean, having a protein content in excess of 50 percent. It is used in doughnuts, cereal, bread, and sausage products for protein fortification and binding.

Soybean Oil The oil obtained from the seed of the soybean legume. It consists of approximately 86 percent unsaturated fatty acids with linoleic and oleic being the principal two fatty acids. It exists in hydrogenated and unhydrogenated forms. It is used in shortenings and margarine in the hydrogenated form. It has some use in salad and cooking oils in the unhydrogenated form, but is limited by its tendency to develop undesirable odor and flavor when in contact with air or when heated to frying temperatures.

Soybean Protein The protein obtained from soybeans, containing the essential amino acids. The most common forms are soybean flour (approximately 50 percent protein), soybean concentrate (approximately 70 percent protein), and soybean protein isolate (approximately 90 percent protein). It is used in sausages, snack foods, and meat analogs to provide emulsification, binding, moisture control, texture control, and protein fortification.

Soybean Protein Concentrate The concentrate obtained by processing soybean flour to remove the soluble carbohydrates. The protein content is approximately 70 percent. In the powder form, it is used in processed meat products and sausage products for moisture and fat binding as well as texture. In baby food, cereal, and snack food it provides protein fortification. In the granular form, it is used in ground meat food items for texture.

Soybean Protein Isolate The isolate prepared from soybean flour by extracting the protein and precipitating it to yield a product of approximately 90 percent protein. It functions to increase the protein content in foods, to reduce shrinkage, and to provide structure and appearance by emulsifying, stabilizing, and binding the fat and water. It is used in frozen spaghetti and meatballs, whipped toppings, and snack foods.

Spice   A variety of dried plant products that exhibit an aroma and flavor and from which no volatiles or other flavoring principles have been removed. See **Spice category**.

Spirit Vinegar   The product made by the acetous fermentation of dilute distilled alcohol, containing not less than 4 g of acetic acid per 100 cm^3 at 20°C. It functions as an acidulant and provider of flavor. It is used in mayonnaise, sauces, and salad dressings. It is also termed distilled vinegar and grain vinegar.

Stannous Chloride $SnCl_2 \cdot 2H_2O$. An antioxidant and preservative that exists as white or colorless crystals, being very soluble in water. It reacts readily with oxygen, preventing its combination with chemicals and foods which would otherwise result in discoloration and undesirable odors. It is used for color retention in asparagus at less than 20 parts per million. It is also used in carbonated drinks.

Starch   A carbohydrate consisting of glucose units containing amylose and amylopectin which contribute to varying starch properties. Starch is insoluble in cold water, but upon heating the starch granules swell and burst forming starch paste. Starch sources include arrowroot, corn, potato, rice, sago, tapioca, waxy corn, and wheat. Starches are modified by treatment to alter their functional properties. Terminology designating these starches includes acid-modified corn starch, food starch modified, modified food starch, oxidized corn starch, pregelatinized starch, thin-boiling starch, and wheat starch, gelatinized. See **Specific Starch**.

Stearic Acid $C_{18}H_{36}O_2$. A fatty acid that is a mixture of solid organic acids obtained principally from stearic acid and palmitic acid. It is practically insoluble in water. It functions as a lubricant, binder, and defoamer. It is used as a softener in chewing gum base.

Stearoyl Lactylate   A dough conditioner, emulsifier, and whipping agent that is the reaction product of stearic and lactic acid neutralized to a sodium or calcium salt, for example, calcium stearoyl lactylate and sodium stearoyl lactylate. It is used to improve the tolerance of bread dough to processing and to improve gas retention. It is used as an emulsifier in coffee whiteners, puddings, and lowfat margarine. It functions as a whipping aid in egg products and vegetable fat toppings. It complexes starch in dehydrated potatoes to allow for production of thicker, more uniform sheets.

Stearoyl Propylene Glycol Hydrogen Succinate See **Succistearin**.

Stearyl Citrate   An antioxidant made by reacting citric acid (not soluble in fats and oils) with stearyl alcohol which readily dissolves in oils, thus enabling the citrate to dissolve in oil. It prevents metal ions from catalyzing oxidative reactions which cause rancidity. It is related to isopropyl citrate. It is used in vegetable oils and margarine.

Sterculia Gum   See **Karaya**.

Straight Flour   All of the flour that can be milled from a wheat blend, or approximately 72 percent of the wheat kernel which equates to 100 percent separation. See **Flour category**.

Succinylated Monoglycerides   Emulsifiers and dough conditioners made by the dissociation of succinylated monoglycerides. They are used in baked goods at 0.056 to 0.113 kg per 45.4 kg of flour to provide dough strength, improve shelf life, and improve texture. They are also used in shortenings.

Succinic Acid $C_4H_6O_4$. An acidulant that is commercially prepared by the hydrogenation of maleic or fumaric acid. It is a nonhygroscopic acid but is more soluble in 25°C water than fumaric and adipic acid. It has low acid strength and slow taste build-up; it is not a substitute for normal acidulants. It combines with proteins in modifying the plasticity of bread dough. It functions as an acidulant and flavor enhancer in relishes, beverages, and hot sausages.

Succinic Anhydride $C_4H_4O_3$. An acidulant that hydrolyzes very slowly to succinic acid in water. It has thermal stability and a low melting point (118°C) which permits it to be used in products at comparatively low temperatures. It is used as a leavening acidulant for baking powder.

Succistearin   An emulsifier which is also termed stearoyl propylene glycol hydrogen succinate. It is used in shortening to help improve the tenderness in baked goods.

Sucrose   See **Sugar**.

Sugar   A sweetener that is the disaccharide sucrose, consisting of one molecule of glucose and one molecule of fructose. It is obtained as cane or beet sugar. It has relatively constant solubility and is a universal sweetener because of its intense sweetness and solubility. It is available in various forms which include granulated, brown, and powdered. It is used in desserts, beverages, cakes, ice cream, icings, cereals, and baked goods. It is also termed beet sugar, cane sugar, and sucrose.

Sugar, Brown   See **Brown Sugar**.

Sugar, Natural   See **Turbinado Sugar**.

Sugar, Powdered   See **Powdered Sugar**.

Sugar, Raw   A natural sugar that has been washed to remove the impurities. It has a light golden color resulting from the molasses and a larger crystal size than granulated sugar. It is used where the flavor of natural sugar is desired, such as in cookies, bread, and cakes.

Sugar, Reducing   See **Reducing Sugar**.

Sugar Syrup   A sweetener that is clear solutions of sucrose existing in varying grades. There is a water-white grade which is a sparkling clear syrup used in canned goods and beverages. There is also a light straw grade which has small amounts of color and nonsugar components.

Sugar Syrup, Invert   See **Invert Sugar Syrup**.

Sugar, Washed Raw   See **Turbinado Sugar**.

Sulfur Dioxide SO_2. A preservative, being a gas that dissolves in water to yield sulfurous acid. Sulfite salts, such as sodium and potassium sulfite, sodium and potassium bisulfite, and sodium and potassium metabisulfite, yield free sulfurous acid at low pH. Sulfur dioxide prevents the discoloration of foods by combining with the sugars and enzymes. It also inhibits bacterial growth. It is used in beverages, cherries, wines, and fruits.

Sulfuric Acid H_2SO_4. An acidulant that is a clear, colorless, odorless liquid with great affinity for water. It is prepared by reacting sulfur dioxide with oxygen and mixing the resulting sulfur trioxide with water, or by reacting nitric oxide with sulfur dioxide in water. It is very corrosive. It is used as a modifier of food starch and is used in caramel production and in alcoholic beverages.

Sunflower Oil   A highly polyunsaturated oil obtained from sunflower seeds. There are two types of sunflower grown: an oilseed type used as a vegetable oil, and a non-oilseed type used for human food and bird seed. The composition of sunflower oil varies according to location and growing temperature. In general, sunflowers grown above the 39th parallel are high in linoleic acid and those grown below are high in oleic acid. The high linoleic variety is used for margarine and salad oil, while the high oleic variety is used in frying applications. This bland-flavored oil has a smoke point of 485 to 490°F (252–254°C) which gives it utility in baking, cooking, and frying foods. It is also used as a salad oil. In the hydrogenated form, it is used in margarine and shortenings.

Sunset Yellow FCF   See **FD&C Yellow #6**.

Sweet Pepper   See **Paprika**.

Sweet Rice Flour   See **Waxy Rice Flour**.

Tallow Animal fat obtained by separation from connected tissue, usually in mutton or beef. It consists principally of oleic and palmitic acid. It is a source of fat and is used in cake mix. It is used mostly in shortening and cooking oils.

Tannic Acid A sequestrant that refers to a mixture of hydrolyzable tannins of a more complex structure than gallic acid. It is used in clarifying beer and wine. See **Tannins**.

Tannins These are phenolic compounds that have several hydrolyzable groups. They are classified as: (a) hydrolyzable, yielding phenols such as gallic acid in the presence of acid and heat; and (b) condensed, obtained from the extract of oak trees and not hydrolyzable. Tannins are used for taste and chemical properties and as a sequestrant. They affect the color and flavor of fruits and vegetables. They are used in fruits, wine, and beer to remove undesirable material by forming insoluble complexes with the proteins.

Tapioca Starch Starch having a bland flavor, being opaque, and forming long, cohesive pastes. It is found mainly in the modified form, being the pearl, granulated form which has been treated to be less stringy. It is used in puddings and pie fillings.

Taro The tropical tuber *Colocasia esculenta* which can be used to make poi. Poi is a carbohydrate food made by cooking the underground stem (corm) of the taro plant. The corms must be cooked because the calcium oxalate crystals present in the raw vegetable will act as tiny needles in the mouth.

Tarragon   The dried leaves and flowering tops of the herb *Artemisia dracunculus* L. It has a distinct aroma and anise-like flavor. It is used in salads, fish, sauces, and vinegar.

Tartaric Acid $C_4H_6O_6$. An acidulant that occurs naturally in grapes. It is hygroscopic and rapidly soluble, with a solubility of 150 g in 100 ml of distilled water at 25°C. It has a slightly tarter taste than citric acid, with a tartness equivalent of 0.8 to 0.9. It augments the flavor of fruits in which it is a natural constituent. It is used in grape- and lime-flavored beverages and grape-flavored jellies. It is used as an acidulant in baking powder and as a synergist with antioxidants to prevent rancidity.

Tartrazine   See **FD&C Yellow #5**.

Tertiary Butylhydroquinone (TBHQ) $C_{10}H_{14}O_2$. An antioxidant that exhibits an excellent stabilizing effect in unsaturated fats and oils. It has good solubility in fats and oils, with a maximum usage level of 0.02 percent based on the weight of the fat or oil or the fat content of the food product. It shows no discoloration in the presence of iron and produces no discernible flavor or odor. It can be combined with BHA and BHT. It is used in edible fats and vegetable oils to retard rancidity. It is used in potato chips and dry cereal. It is also termed mono-tertiary-butylhydroquinone.

Tetrahydrofurfuryl Acetate   A synthetic flavoring agent that is a stable, colorless liquid of slightly fruity odor. It should be stored in glass or tinned containers. It is used in fruit flavors for application in candy and baked goods at 2 to 20 parts per million.

Tetrahydrofurfuryl Propionate   A synthetic flavoring agent that is a stable, colorless liquid of chocolate note. It should be stored in glass or tin-lined containers. It is used in flavors for chocolate with applications in beverages, and ice cream at 2 parts per million and in candy and baked goods at 20 parts per million.

Tetrapotassium Diphosphate   See **Tetrapotassium Pyrophosphate**.

Tetrapotassium Pyrophosphate $K_4P_2O_7$. A moisture binder, sequestrant, emulsifier, and buffer that is mildly alkaline, with a pH of 10. It is very soluble in water with a solubility of 187 g per 100 ml at 25°C. It is used for moisture binding in meat and is used in low-sodium products. It is also termed tetrapotassium diphosphate, potassium pyrophosphate, and TKPP.

Tetrasodium Diphosphate See **Tetrasodium Pyrophosphate**.

Tetrasodium Pyrophosphate $Na_4P_2O_7$. A coagulant, emulsifier, and sequestrant that is mildly alkaline, with a pH of 10. It is moderately soluble in water, with a solubility of 0.8 g per 100 ml at 25°C. It is used as a coagulant in noncooked instant puddings to provide thickening. It functions in cheese to reduce the meltability and fat separation. It is used as a dispersant in malted milk and chocolate drink powders. It prevents crystal formation in tuna. It is also termed sodium pyrophosphate, tetrasodium diphosphate, and TSPP.

Textured Soy Flour Soy flour that is processed and extruded to form products of specific texture and form, such as meatlike nuggets. The formed products are crunchy in the dry form and upon hydration become moist and chewy.

Textured Vegetable Protein A vegetable protein that is processed and extruded to form beeflike strips, meatlike nuggets, or other analogs. In the dehydrated form, the analogs are crunchy and upon hydration become moist and chewy. Soy protein is the most popular protein source, but other vegetable proteins include peanut and wheat. It is used as meat analogs. It is also termed textured soy flour or textured soy protein.

Thiamine $C_{12}H_{17}ClN_4O_5$. The water-soluble vitamin B_1, required for normal digestion and functioning of nerve tissues and in the prevention of beriberi. It also acts as a coenzyme in the metabolism of carbohydrates. During processing, the higher and longer the heating period, the greater the loss. The loss is reduced in the presence of acid. Thiamine hydrochloride and thiamine mononitrate are two available forms. The mononitrate form is less hygroscopic and more stable than the hydrochloride form, making it suitable for use in beverage powders. It is used in enriched flour and is found as thiamine mononitrite ($C_{12}H_{17}N_5O_4S$) in frozen egg substitute and crackers.

Thiamine Mononitrate   See **Thiamine**.

Thin-Boiling Starch   See **Cornstarch, Acid-Modified**.

Thiodipropionic Acid $C_6H_{10}O_4S$. An antioxidant used to prevent fats and oils from going rancid. It has the same functionality as BHA, BHT, and propyl gallate.

Thyme   The dried leaves and flowering tops of the shrub *Thymus vulgaris* L. There are two important variations: French thyme, which has a narrow leaf; and lemon thyme, which has a variegated leaf. It is used in soups, cheese, sauces, and appetizers.

Titanium Dioxide TiO_2. A white pigment that disperses in liquids and possesses great opacifying power. The crystalline modifications of titanium dioxide are rutile and anatase, of which only anatase finds use as a color additive. It is used in confectionary panned goods, icings, and sugar syrups for candy.

Tocopherol   Fat-soluble vitamin E, which is a light yellow oil readily degradable by heat. As a vitamin, it is essential for normal muscle growth and prevents vitamin A destruction by deterioration. It also functions as an antioxidant. It prevents the oxidation of certain fatty acids and is stable unless the food becomes rancid. Vegetable oils contain a higher concentration of natural antioxidants, including tocopherols, than animal fats and are thus more stable. Tocopherol is obtained from vegetable oils, beans, eggs, and milk.

Tofu   A soybean curd product. Soybeans are soaked, ground, and filtered, with the remaining mixture being heated to 75°C and a coagulant added, which results in the formation of the soy curd and whey. The soy curd is pressed to separate it from the whey and then washed and cooled. It is low in calories and saturated fats while high in vitamins, minerals, and digestible protein. It is tasteless, but takes on the flavors of the products with which it is cooked. Uses include frozen desserts and meat products.

Tomato Paste   The paste prepared from tomatoes which are processed by heat to prevent spoilage. The paste contains not less than 24 percent tomato soluble solids.

Tragacanth   A gum produced from a bush of the genus *Astragalus*. It swells in water to give a highly viscous sol or paste. A 1 percent solution of the purest gum has a viscosity of approximately 3400 cps, and about 2 percent can form a paste. Solutions have a pH of 5 to 6. It is stable at low pH and is an effective suspending agent because of its stability and acid resistance. It is used in salad dressings, sauces, fruit fillings, and citrus beverages.

Triacetin See **Glyceryl Triacetate**.

Tribasic Calcium Phosphate See **Tricalcium Phosphate**.

Tricalcium Orthophosphate See **Tricalcium Phosphate**.

Tricalcium Phosphate $Ca_3(PO_4)_2 \cdot H_2O$. An anticaking agent and calcium source that is a white powder that is almost insoluble in water. It is used as an anticaking agent in table salt and dry vinegar. It is used as a source of calcium and phosphorus in cereals and desserts. It functions as a bleaching agent in flour and in lard, and prevents undesirable coloring and improves stability for frying. It is also termed tribasic calcium phosphate, tricalcium orthophosphate, and precipitated calcium phosphate.

Tricalcium Silicate An anticaking agent used in table salt.

Triethyl Citrate $C_{12}H_{20}O_7$. A sequestrant that is an oily liquid, slightly soluble in water. It is found in lemon drinks.

Trihydroxybutyrophenone (THBP) An antioxidant.

Tripotassium Citrate $C_6H_5K_3O_7 \cdot H_2O$. A buffer and sequestrant that possesses the advantageous properties of citric acid but without the acid reactions. See **Potassium Citrate, Monohydrate**.

Tripotassium Monophosphate See **Tripotassium Phosphate**.

Tripotassium Orthophosphate See **Tripotassium Phosphate**.

Tripotassium Phosphate An emulsifier and alkaline buffer that is strongly alkaline, with a pH of 12. It is very soluble in water, with a solubility of 107 g per 100 ml at 25°C. It is insoluble in alcohol. It is used in low-sodium products. It is also termed tripotassium orthophosphate, potassium phosphate tribasic, tripotassium monophosphate, and TKP.

Trisodium Citrate A buffer and sequestrant that is the trisodium salt of citric acid. See **Sodium Citrate**.

Trisodium Monophosphate   See **Trisodium Phosphate**.

Trisodium Orthophosphate   See **Trisodium Phosphate**.

Trisodium Phosphate Na_3PO_4. An emulsifier and buffer that is strongly alkaline, with a pH of 12. It is moderately soluble in water, with a solubility of 14 g per 100 ml at 25°C. It functions as an emulsifier in processed cheese to improve texture. It maintains viscosity and prevents phase separation in evaporated milk and is also found in cereals. It is also termed trisodium orthophosphate, sodium phosphate tribasic, and trisodium monophosphate.

Trisodium Phosphate Crystals $Na_3PO_4 \cdot 12H_2O \cdot 1/4NaOH$. An emulsifier and buffer with a solubility in water of 50 g per 100 ml at 25°C. It is used in processed cheese as an emulsifier and it is also used in denture cleaner formulations.

Turbinado Sugar   Washed raw sugar of light gold color and larger grain size than regular sugar. It has a thin film of molasses which contributes toward the distinctive flavor. It is also termed natural sugar and washed raw sugar.

Turmeric   A spice and colorant that is the rhizome or root of *Curcuma longa*. As a spice, it has a taste related to mustard. As a vegetable color, it has a bright yellow to greenish-yellow hue. The yellow pigment is curcumin. It is water miscible and has excellent heat stability, poor light and pH stability, and good tinctorial strength. It exists as an extract and oleoresin. It is used in meat, poultry, fish, and rice dishes.

(Gamma)-Undecalactone $C_{11}H_{20}O_2$. A synthetic flavoring agent that is a colorless to yellow liquid of strong peach fruit odor. It is unstable to alkali and stable to weak organic acids. It should be stored in glass or tin containers. It is used in flavors for its peach note and has application in gelatins, puddings, beverages, ice cream, and candy at 7 to 11 parts per million.

Vanaspati A vegetable fat used in candy.

Vanilla A flavorant obtained from the cured vanilla bean. The vanilla or vanilla bean refers to the fully grown, unripe, cured and dried fruit pod of the vanilla vine *Vanilla planifolia*. Those beans produced in Madagascar and its neighboring islands are termed "Bourbon beans"; those produced in Indonesia are termed "Java beans." The bean contains 1.5 to 3.0 percent vanillin, the most powerful flavorant in the cured bean, along with approximately 10 percent of other extractives. It is used in the comminuted form in "Philadelphia" type ice cream or as a vanilla flavorant in sauces or liquids by suspending the whole bean in them. Most vanilla flavoring is done with vanilla extract.

Vanilla Extract A flavorant made from vanilla bean extract. It is a solution containing not less than 35 percent alcohol of the components extracted from one or more units of vanilla constituent. One unit is 0.378 kg of vanilla beans containing not more than 25 percent moisture. A double-strength solution (2-fold) contains twice the quantity of beans. It is used in desserts, baked goods, and beverages.

Vanilla Flavor, Artificial A flavorant that consists of vanilla reinforced with synthetic vanillin. The best imitation vanillas contain vanillin, ethyl vanillin, or very little coumarin with or without vanilla, while the poorer ones contain high levels of coumarin. It is used in desserts, baked goods, and beverages.

Vanilla Sugar   A flavorant consisting of sugar mixed with vanilla extract. It is used in desserts and other food products.

Vanillin $C_8H_8O_3$. A flavorant made from synthetic or artificial vanilla which can be derived from lignin of whey sulfite liquors and is synthetically processed from guaiacol and eugenol. The related product, ethyl vanillin, has 3.5 times the flavoring power of vanillin. Vanillin also refers to the primary flavor ingredient in vanilla, which is obtained by extraction from the vanilla bean. Vanillin is used as a substitute for vanilla extract, with application in ice cream, desserts, baked goods, and beverages at 60 to 220 parts per million.

Vanillin Acetate   A synthetic flavoring agent that is moderately stable, white to yellow crystals of vanilla odor. It should be stored in glass or polyethylene-lined containers. It is used in flavors for vanilla note, with application in beverages, ice cream, candy, and baked goods at 11 to 28 parts per million.

Vegetable Gums   Gums that are water thickeners obtained from a plant source. See **Gum category**.

Vegetable Oils   Oils obtained from a vegetable source, including soy beans, peanuts, cottonseeds, and palms. They are used in cooking and salad oils and dressings.

Vegetable Oil, Hydrogenated   See **Hydrogenated Vegetable Oil**.

Vegetable Protein, Textured   See **Textured Vegetable Protein**.

Vinegar   An acidulant and flavorant that, with regard to general types, is the product produced from cider, grapes, sucrose, glucose, or malt by successive alcoholic and acetous fermentations in which acetic acid is the principal measured component. The term *vinegar* applies only to cider vinegar, also termed apple vinegar. The acetic acid content is measured in grains, where 10 grains equals 1 percent acetic acid. It is used in salad dressing, dressings, and sauces.

Vinegar, Distilled   The product made by the acetous fermentation of dilute distilled alcohol without addition of color, containing not less than 4 g of acetic acid per 100 cm^3 at 20°C. The acetic acid content is measured in grains, where 10 grains equal 1 percent acetic acid. It is used in mayonnaise, salad dressing. It is also termed spirit vinegar and grain vinegar.

Vital Wheat Gluten A powder of high protein content obtained by drying freshly washed gluten under controlled-temperature conditions. It absorbs approximately twice its weight of water and readily forms a cohesive, elastic dough. It is used in bread, rolls, and buns. See **Wheat Gluten**.

Vitamin A See **Retinol**.

Vitamin B₁ See **Thiamine**.

Vitamin B₂ See **Riboflavin**.

Vitamin B₅ See **Pantothenic Acid**.

Vitamin B₆ See **Pyridoxine**.

Vitamin B₆ Hydrochloride See **Pyridoxine Hydrochloride**.

Vitamin B₁₂ See **Cyanocobalamin**.

Vitamin C See **Ascorbic Acid**.

Vitamin D₂ See **Calciferol**.

Vitamin E See **Tocopherol**.

Vitamin K A fat-soluble vitamin that is essential for blood clotting. It is destroyed by irradiation during processing but has no appreciable loss during storage. It occurs in spinach, cabbage, liver, and wheat bran.

Washed Raw Sugar A raw sugar of larger grain size. See **Turbinado Sugar**.

Water H_2O. A colorless, odorless, tasteless liquid formed by the combination of two hydrogen and one oxygen atoms. It allows substances to dissolve and functions as a solvent, dispersing medium, hydrate, and promoter of chemical changes. It is a major constituent in meats, fruits, and vegetables. Distilled water is obtained by condensation of water vapor.

Waxy Corn A corn consisting essentially of amylopectin (pure branched-chain polymers), which differentiates it from regular corn, which consists of amylose and amylopectin. The amylopectin content results in a starch which upon heating forms a clear, cohesive paste that does not form a true gel upon cooling. It has a high water-binding capacity and is resistant to gel formation and retrogradation. It is used in puddings and sauces.

Waxy Maize Starch The starch portion of waxy corn, consisting essentially of amylopectin. It yields pastes that are almost clear upon cooling and are noncongealing. It forms a translucent, water-soluble coating when dried in thin films. It is used to thicken a variety of foods such as sauces and puddings. It is also termed waxy starch and amioca.

Waxy Rice Flour A flour obtained from waxy rice, which contains almost no amylose. It is comparable in viscosity characteristics to waxy corn flour. It has less than 0.5 percent amylose in the starch and contains alpha-amylose. It has excellent resistance to syneresis during freeze/thaw cycles. It is used in frozen sauces and gravies. It is also called sweet rice flour.

Waxy Rice Starch   A rice starch that contributes freeze-thaw stability to sauces and puddings but may provide objectionable flavor.

Waxy Sorghum   A type of sorghum characterized by having paste clarity, high water-binding capacity, and resistance to gel formation and retrogradation. The unmodified form results in a stringy, cohesive paste. It is used in dressings with other starches.

Waxy Starch   See **Waxy Maize Starch**.

Wheat   A cereal grain in which the kernel is separated by milling into flour, bran, and germ. It is used in all types of farinaceous foods. See **Flour category**.

Wheat Flour   A fine powdery substance obtained by milling wheat with application in farinaceous foods. See **Flour category**.

Wheat Germ   The oil-containing portion of the wheat kernel.

Wheat Gluten   The water-insoluble complex protein fraction separated from wheat flours. Gum gluten is wheat gluten in its freshly extracted wet form. Dry gluten is approximately 70 to 80 percent protein, but is deficient in the amino acid lysine. It absorbs two to three times its weight in water. The differences in properties of wheat gluten in comparison to almost all other food proteins are largely due to the low polarity level of the total amino acid structure. Most food proteins have polar group levels of 30 to 45 percent and have a net negative charge, while wheat gluten has a polar group level of approximately 10 percent with a net positive charge. This results in the repulsion of excess water and the close association of the wheat gluten molecules and resistance to dispersion. In baked goods, this results in the ability to form adhesive, cohesive masses, films, and three-dimensional networks. Gluten formation is utilized in the baking industry to impart dough strength, gas retention, structure, water absorption, and retention with breads, cakes, doughnuts, and so on. It is also used as a formulation aid, binder, filler, and tableting aid. See **Gluten**.

Wheat Starch   A starch obtained from wheat. It produces lower viscosity and more tender gels than starch obtained from corn or sorghum. It has a gelatinization range about 10°C lower than corn or waxy maize starch. It is used in the baking industry to permit the use of hard wheat flour in baked goods. It functions as a binder in breading and batter mixes. It is used in soups, pie fillings, sauces, and gravies.

Wheat Starch, Gelatinized   See **Pregelatinized Starch**.

Whey   The portion of milk remaining after coagulation and removal of curd. There are two principal types: sweet whey obtained during the making of rennet-type hard cheeses like Cheddar and Swiss, with a pH of approximately 6.1; and acid whey obtained during the making of acid-type cheeses, such as cottage cheese, with a pH of approximately 4.4 to 4.6. Whey is used as a source of lactose, milk solids, and whey proteins. It is used in baked goods, ice cream, and dry mixes.

Whey Solids   The solid fraction or dry form of whey. It is used as a replacement for milk solids–not-fat to provide a source of protein, solids, and flavor. It is used in baked goods, ice cream, dry mixes, and beverages.

Whole Milk   See **Milk**.

Whole Milk Solids   The product resulting from the drying or desiccation of milk. It contains not less than 26 percent fat and not more than 5 percent moisture. It is used in dry mixes such as puddings. It is also termed dried milk and milk powder.

Whole Wheat Flour   The flour obtained by grinding cleaned wheat, other than durum wheat or red durum wheat, with the proportions of the natural constituents, other than moisture, remaining unaltered. The moisture content is not more than 15 percent. Optional ingredients include malted wheat, wheat flour, and barley flour for compensation for any natural deficiency of enzymes; ascorbic acid; and bleaching ingredients. It is also termed graham flour and entire wheat flour.

Wine Vinegar   The vinegar made by the alcoholic and acetous fermentation of the juices of grapes or wine. It contains a minimum of 4 g per 100 cm^3 acid expressed as acetic acid. There is red wine vinegar, which has a rose to deep red color, and white wine vinegar, which has a pale yellow to off-white color. It is used in salad dressings, marinades, and sauces.

Wintergreen Oil   See **Methyl Salicylate**.

Worcestershire Sauce   A sauce consisting of water, vinegar, soy sauce, corn syrup, salt, spices, or variations of these ingredients. It is used as a flavorant and is found in barbeque sauce and sweet-and-sour type sauces.

Wormwood   See **Artemisia**.

Xanthan Gum A gum obtained by microbial fermentation from the *Xanthomonas campestris* organism. It is very stable to viscosity change over varying temperatures, pH, and salt concentrations. It is also very pseudoplastic which results in a decrease in viscosity with increasing shear. It reacts synergistically with guar gum to provide an increase in viscosity and with carob gum to provide an increase in viscosity or gel formation. It is used in salad dressings, sauces, desserts, baked goods, and beverages at 0.05 to 0.50 percent.

Xylitol $C_5H_{12}O_5$. A sweetener that is a natural sugar substitute commercially made from the polysaccharide xylan obtained from birch trees. It is as sweet as sucrose and has a negative heat of solution which results in a cooling effect. It has a lower viscosity than sugar. It is used in chewing gum, throat lozenges, and chocolate.

Yeast A leavening and fermentation agent that is a single-celled plant that can convert sugar to carbon dioxide. It is used as a leavening agent in bread and dough-type mixtures. It provides a yeasty flavor and tender crust. It has slow action as a leavening agent. One pound of active dry yeast replaces approximately 2 pounds of fresh yeast. Selected yeast strains are used in wine fermentation.

Yeast Extract A flavor contributor and flavor enhancer consisting of a combination of nucleic acids, peptides, polypeptides, amino acids, and other constituents. It is obtained from the yeast cells of *Saccharomyces cerevisiae*, formed in the brewing of beer. It is used to provide the same functions as monosodium glutamate, although not to the same extent. It is used as a partial substitute for meat extract and also functions with other flavor ingredients such as hydrolyzed vegetable proteins. It is used in soups, gravies, spreads, dressings, and meat products. Typical usage levels range from 0.1 to 0.5 percent.

Yeast Food A complete food used in doughs. It contains dough conditioner ingredients such as calcium salts, sulfates, and phosphates which strengthen the gluten. It also contains ammonium salts and phosphates which function as yeast nutrients. It is used in bread dough and in the fermentation of alcoholic beverages.

Yellow Prussiate of Soda $Na_4Fe(CN)_6 \cdot 10H_2O$. An anticaking agent and crystallizing agent. It is sometimes added as a crystallizing agent to salt when it crystallizes to generate jagged and bulky crystals which resist caking. It also functions as a water-soluble anticaking agent. It is also termed sodium ferrocyanide.

Yogurt   A custard-like or soft gel product made by fermenting milk with bacterial cultures, specifically *Lactobacillus bulgaricus* and *Streptococcus thermophilus*, to a pH range of usually 4.0 to 4.5. It is used as a snack; as a meal; or in desserts, salad dressings, and baked goods.

Yucca Plant Extract   A foaming agent obtained from the yucca plant species *Yucca brevifolia* and *Yucca schidigera*. It is available in liquid concentrate or dried form, is dark brown, and has a slight bittersweet flavor, and a pH of 4.0. It is stable over a wide pH range and heat treatment and is readily soluble in water. It is used in applications where a frothy appearance and foam stability are desired, such as in root beer, cocktail mixes, and whipped beverages. Usage level is 50 to 150 parts per million.

Zein A corn protein produced from corn gluten meal. It lacks the amino acids lysine and tryptophan, so it is not suitable as a sole source of dietary protein. It is insoluble in water and alcohols but is soluble in aqueous alcohols, glycols, and glycol ethers. It functions as a film and coating to provide a moisture barrier for nuts and grain products. It also functions as a coating for confections and a glaze for panned goods.

Zinc Zn. A metallic element that functions as a nutrient and dietary supplement. It is believed to be necessary for nucleic acid metabolism, protein synthesis, and cell growth. Sources of zinc include zinc acetate, carbonate, chloride, gluconate, stearate, and sulfate.

Zinc Acetate A source of zinc that functions as a nutrient and dietary supplement.

Zinc Carbonate A source of zinc that functions as a nutrient and dietary supplement.

Zinc Chloride A source of zinc that functions as a nutrient and dietary supplement.

Zinc Gluconate $C_{12}H_{22}ZnO_{14}$. A source of zinc that functions as a nutrient and dietary supplement.

Zinc Oxide A source of zinc that functions as a nutrient and dietary supplement.

Zinc Stearate   A source of zinc that functions as a nutrient and dietary supplement.

Zinc Sulfate   A source of zinc that functions as a nutrient and dietary supplement. It exists as prisms, needles, or powder. It has a solubility of 1 g in 0.6 ml of water. It is found in frozen egg substitutes.

Ingredient Categories

Acidulants

Acidulants are acids used in processed foods for a variety of functions that enhance the food. Acids are used as flavoring agents, preservatives in microbial control, chelating agents, buffers, gelling and coagulating agents, and in many other ways. Examples of these functions are:

1. Flavoring agent — contributes and enhances flavor in carbonated beverages, fruit drinks, and desserts.
2. Preservative — an acid medium restricts the growth of spoilage organisms in mayonnaise, tomato sauce; and retards the activity of enzymes involved in discoloration in fruits.
3. Chelating agent — aids in binding metals that can cause oxidation in fats and oils, and discoloration in canned shrimp.
4. Buffer — maintains and controls acidity during processing, and maintains acidity within a given range in prepared desserts.
5. Gelling agent — controls the gelling mechanism of algin and pectin gels such as desserts and jams.
6. Coagulating agent — reduction of pH results in coagulation of milk protein which is used in preparation of direct acidified cheese and desserts.

Acidulant selection depends upon the application or processed food. The properties to consider are flavor profile, pH, solubility rate, solubility, and hygroscopicity. Flavor profile refers to the perceived sharpness or blandness contributed by the acid. Tartaric and citric acids provide a sharp taste as compared to lactic acid, which provides a blander taste.

The acids provide different pHs at similar concentrations, falling generally in the range of pH 2 to 3 at 1 percent concentration. The solubility rate determines how rapidly the acid dissolves and contributes toward the flavor profile. In a beverage powder dissolved in cold water, rapid solubility is required, so perhaps citric acid would be used. The solubility of the acid refers to the quantity in solution. Cream of tartar and fumaric acid have low solubilities, which makes them suitable for bakery applications in controlling leavening systems. Phosphoric acid, a liquid, is miscible with water and used in beverage syrups. Hygroscopic acids will absorb moisture, and hygroscipicity needs to be considered when the application is dry mixes. If hygroscopic acids, such as citric or tartaric acids, are used in dry mixes, proper packaging is essential. Alternative less hygroscopic acids are adipic and fumaric acid.

Table 1 illustrates the properties of the principal acids used in the food industry.

Antioxidants

Antioxidants are chemical compounds that provide stability to fats and oils by delaying oxidation, (which involves the loss of electrons, and the gain of oxygen). The oxidation of fats and oils is believed to occur as a series of chain reactions in which oxygen from the air is added to the free fat radical. The fat molecule loses a hydrogen atom and becomes an unstable free radical with a high affinity for oxygen. Oxygen is added and the fat molecule, to complete its electron structure, reacts with another fat molecule and removes a hydrogen atom. This produces another free radical and results in a chain reaction. The antioxidant functions by replacing the fat molecule as the hydrogen atom donor in order to complete the electron structure of the free radical, thus terminating the chain reaction. Thus oxidative rancidity, which results in off-flavors and odors, is retarded until the antioxidant supply is used.

The most commonly used antioxidant formulations contain combinations of BHA (butylated hydroxyanisole), BHT (butylated hydroxytoluene), and propyl gallate. These formulations usually contain a chelating agent (reacts with metal to form a complex and thus prevents the metal from acting as a catalyst in oxidative reactions), of which citric acid is the most common. Natural antioxidants such as the tocopherols and guaiac gum usually lack the potency of BHA, BHT, and propyl gallate combinations.

Antioxidants are effective at low concentrations, that is, 0.02 percent or less, based on the fat or oil content of the food. Examples of applications are:

Table 1. COMPARATIVE ACID CHART

Name	Chemical Formula	Solubility (g/100 ml Distilled H_2O at 20°C)	pH in H_2O Percentage Solution	pH	Tartness	Comments
Adipic acid	$C_6H_{10}O_4$	1.9	0.6 1.2	2.86	Smooth tart	Nonhygroscopic; imparts long-lasting flavor note
Citric acid, anhydrous	$C_6H_8O_7$	146	0.12 0.5	3.20 2.35	Sharp tart	Immediate acid taste
Citric acid, monohydrate	$C_6H_8O_7 \cdot H_{20}$	175	—	—	Sharp tart	Immediate acid taste
Fumaric acid	$C_4H_4O_4$	0.49	0.5	2.15	Tart	Nonhygroscopic; 3 lb replaces 5 lb citric acid to produce taste equivalent
Gluconic acid	$C_6H_{12}O_7$	100	1	2.80	Mild	Unobtrusive
Glucono-Delta-Lactone	$C_6H_{10}O_6$	59	See gluconic acid		Mild	Low hygroscopicity; slowly converted to gluconic acid
Lactic acid	$C_3H_6O_3$	Liquid	5	2.25	Smooth tart	Slow onset of acid taste which lingers
Malic acid	$C_4H_6O_5$	130	1	2.35	Smooth tart	Hygroscopic; perception of sourness or sharpness, starts gradually, rises to peak and fades slowly
Phosphoric acid	H_3PO_4	Liquid	0.12	2.68	—	Sour taste, high acidity
Succinic acid	$C_4H_6O_4$	5	Low acid strength		Tart	Nonhygroscopic; has slow flavor build-up

1. Rendered animal fat, such as lard;
2. Vegetable oils, such as cottonseed oil and corn oil;
3. Food products of high fat content, such as doughnuts and potato chips; and
4. Food products of low fat content, such as cereals and dehydrated potatoes.

Antioxidants should be added to fats and oils before oxidation has started in order to be effective. The antioxidant cannot reverse the oxidation process nor regenerate a product that has become rancid. The oxidation process is accelerated by heat, light, moisture, metals, and other factors.

Antioxidants include trihydroxybutyrophenone (THBP), dilauryl thiodipropionate (DLTDP), nordihydroguaiaretic acid (NDGA), guaiac gum, thiodipropionic acid, tocopherols, lecithin, sodium erythorbate, ascorbic acid, and ascorbyl palmitate.

Chelating Agents (Sequestrants)

Chelation is an equilibrium reaction between a chelating (complexing) agent and a metal ion which forms a complex. Trace metal ions in foods can produce undesirable effects such as discoloration, turbidity, and oxidation. The chelating agents can form a complex with the unwanted trace metals, thus blocking the reactive sites of the metal ions and rendering them inactive. The complex formed is termed a *chelate*, that is, metal + chelating agent = metal complex.

An equilibrium constant K defines the ratio of chelated metal to unchelated metal. The log K is the stability constant, measuring the affinity of the complexing agent for the metal ion. A high K value indicates a high affinity of the complexing agent for the metal ion and thus a low value for free metal ion concentration. Stability constants for some metal ions are shown in Table 2.

The most problematic metal ions in foods are iron and copper. In a system containing several metal ions, the ones with the highest stability constants will be chelated first, followed in order of highest stability constant until the chelating agent is used.

Chelating agents are used to control the reactions of trace metals in foods to principally prevent discoloration, such as occurs in potatoes when iron reacts with phenolic compounds in the presence of oxygen. They are also used with antioxidants to complex trace metals, thus preventing the metal from acting as a catalyst in oxidative reactions. Application examples are the use of:

Table 2. METAL ION STABILITY CONSTANTS

Metal	Log K
Fe^{3+}	25.70
Cu^{2+}	18.80
Ni^{2+}	18.56
Zn^{2+}	16.50
Co^{2+}	16.21
Fe^{2+}	14.30
Mn^{2+}	13.56
Ca^{2+}	10.70
Mg^{2+}	8.69

1. Phosphates in soft drinks to chelate heavy metal ions that interfere with carbonation;
2. EDTA in mayonnaise to eliminate the oxidative activities of trace metals and protect flavor;
3. Sodium acid pyrophosphate to prevent discoloration in potatoes; and
4. Sodium hexametaphosphate to sequester calcium ions and permit the solubilization of alginates.

Examples of chelating agents include calcium disodium EDTA, disodium dihydrogen EDTA, tetrasodium pyrophosphate, citric acid, monoisopropyl citrate, phosphoric acid, and monoglyceride phosphate.

Colors

Colors are usually designated artificial or natural, which indicates that they are, respectively, synthetically manufactured or obtained from natural sources. Synthetic color additives "certified" by the Food and Drug Administration are designated FD&C (Food, Drug, and Cosmetic) and are traditionally termed primary colors. These colors are shades of red, yellow, blue, and green. Secondary colors are blends of certified primary colors, with or without diluents. Table 3 lists the physical and chemical properties of the certified food colors.

Those acceptable food colors not designated "certified" are designated "approved" and consist of natural organic and synthetic inorganic colorants used in certain applications (see Table 4). Colors are available in powders, liquids, granules, pastes, and other forms. Colorant determi-

Table 3. PHYSICAL AND CHEMICAL PROPERTIES OF CERTIFIED FOOD COLORS

FDA Name (Chemical Class)	Hue Range	Tinctorial Strength	Stability to:			Compatability with Food Components
			Light	Oxidation	pH Change	
Red #2 (amaranth)	Bluish-red	Good	Moderate	Fair	Good	Good
Red #3 (erythrosine)	Bluish-pink	Very good	Fair	Fair	Poor	Poor
Red #40	Yellowish-red	Very good	Very good	Fair	Good	Very good
Yellow #5 (tartrazine)	Lemon yellow	Good	Moderate	Fair	Good	Moderate
Yellow #6 (sunset yellow FCF)	Reddish	Good	Moderate	Fair	Good	Moderate
Green #3 (fast green FCFF)	Bluish-green	Excellent	Fair	Poor	Good	Good
Blue #1 (brilliant blue FCF)	Greenish-blue	Excellent	Fair	Poor	Good (unstable in alkali)	Good
Blue #2 (indigotine)	Deep blue	Poor	Very poor	Poor	Poor	Very poor

Source: Food Colors (National Academy of Sciences, 1971)

Table 4. PHYSICAL AND CHEMICAL PROPERTIES OF SOME NONCERTIFIED COLORS

FDA Name (Chemical Class)	Hue Range	Tinctorial Strength	Stability to:				Compatability with Food Components
			Light	Oxidation	pH Change	Microbial Attack	
Annatto extract (carotenoid)	Yellow to peach	Good	Moderate	Very good	Very good	—	Very good
Beets, dehydrated (anthocyanin)	Bluish-red <pH 6 Brown >pH 6	Good	Good	Excellent	Good	Good	Excellent
Caramel	Yellowish to tan	Fair	Good	Good	Good	Fair	Good
Beta-carotene (carotenoid)	Yellow to orange	Good	Fair	Poor	Good	Poor	Good
Cochineal extract; carmine	Orange red to wine red	Fair	Good	Good	Poor	Poor	Good
Grape skin extract	Red-blue-green	Poor	Poor	Poor	Poor	Fair	Good
Tumeric extract and oleoresin	Bright yellow to greenish-yellow	Good	Poor	Moderate	Poor	—	Good

Source: Food Colors (National Academy of Sciences, 1971)

nation includes desired hue, water solubility, and stability. The following groupings are used to illustrate different color groups:

Artificial Coloring

1. Water-soluble colors are designated as FD&C, followed by the color name and number designation, for example, FD&C Blue #2. They have a corresponding common name, for example, indigotine. The colors vary in hue, solubility, and other properties, which relates to the intended application. The water-soluble colors include FD&C Blue #1, Blue #2, Green #3, Red #40, Yellow #5, and #6.

2. Water-insoluble colors are termed FD&C aluminum lakes. Lakes are prepared by the absorption of a certified dye on an insoluble substrate, aluminum hydroxide, and as such include the standard colors. Lakes are used to color dry ingredients, increase stability, and reduce color migration. Lakes can be used to color foods with a high oil or fat content, in dry mixes and coated candies, and for other purposes.

Natural Coloring

Natural colors are usually extracted from botanical sources and often contain several pigments and, as such, are not used as direct replacements for FD&C colors. The colors have low tinctorial strength due to a low quantity of pigment present and thus are used at higher levels than FD&C colors. These colors generally have poor stability in that their color and rate of degradation are affected by pH, temperature, and other conditions. Some natural colorings are:

1. Annatto—the pigment bixin found in the coating of the annatto seed. The color hue ranges from yellow to reddish-orange;
2. Turmeric—contains curcumin obtained from turmeric root. The color hue varies from greenish-yellow to yellow-orange;
3. Paprika—produces a red to red-orange color; and
4. Beet—produced from red beets and has a deep reddish-purple color.

Corn Sweeteners

Corn sweeteners are the products made by using mild conversion techniques to produce starch hydrolysates, that is, dextrin, maltodextrin, and corn-syrup solids. These carbohydrates vary according to sugar composition, which accounts for their differences in properties. They are classified in terms of dextrose equivalent (DE), a standard that expresses the level of reducing sugar calculated as dextrose (see Table 5). Complete conversion of cornstarch yields dextrose, also termed corn sugar, which has a DE of 92. When the reaction is stopped at the intermediate stage, corn syrup, consisting of dextrose, maltose, and polysaccharides, is obtained. It can be manufactured to have the desired DE based on desired properties. A common distribution is 28 to 38 DE (low conversion) to 58 to 68 DE (high conversion). Partial hydrolysis of cornstarch commercially produces maltodextrins of DE ranging from 13 to 22 and dextrin, with a DE range of 7 to 12. In high-fructose corn syrup, a fraction of the dextrose has been converted enzymatically to fructose, thus making a sweeter syrup.

Emulsifiers

Emulsifiers are products that function to reduce the surface tension between two immiscible phases at their interface, allowing them to become miscible. The interface can be between two liquids, a liquid and a gas, or a liquid and a solid. Most emulsions involve water and oil or fat as the two immiscible phases, one being dispersed as finite globules in the other. The liquid as globules is referred to as the dispersed or internal phase, while the medium in which they are suspended is the continuous or external phase. There are two types of emulsions depending on the composition of the phases. In an oil-in-water emulsion such as milk and mayonnaise, the water is the external phase and the oil is the internal phase. In a water-in-oil emulsion such as butter, the oil is the external phase and the water is the internal phase. By use of the proper emulsifier, the two phases will mix and separation is prevented or delayed.

Table 5. COMPARISON OF PROPERTIES RELATIVE TO DEXTROSE EQUIVALENT (DE)

DE	Product	Solubility	Sweetness	Viscosity	Bodying Agent
0	Starch	4	4	1	1
6–20	Maltodextrin	3	3	2	2
20–58	Corn Syrup	2	2	3	3
100	Corn Sugar (Dextrose)	1	1	4	4

Range: 1 = greatest; 4 = least.

The emulsifier consists of a hydrocarbon chain which has affinity for fats and oils (lipophilic group) and a polar group which is attracted to water or aqueous solutions (hydrophilic group). The emulsifier tends to concentrate at the interface between two immiscible liquids, with the hydrophilic portion in the water and the lipophilic portion in the oil. In such fashion, the surface properties are altered by the orientation of the emulsifiers at the interface which reduces the resistance of the two substances to combine. They are frequently used as blends for obtaining the most stable emulsion system. Emulsifiers have the following major functions:

1. Complexing—reaction with starch in bakery products which retards the crystallization of the starch, thus retarding the firming of the crumb which is associated with staling;
2. Dispersing—the reduction of interfacial tension which creates an intimate mixture of two liquids that normally are immiscible, example being oil-in-water emulsions such as salad dressing;
3. Crystallization control—control of crystallization in sugar and fat systems, i.e., chocolate where it allows for brighter initial gloss and prevention of solidified fat on the surface;
4. Wetting—allows the surface to be more attracted to water, such as powders, i.e., coffee whitener in which the addition of surfactant aids the dispersion of the powder in the liquid without lumping on the surface; and
5. Lubricating—functions as a lubricant, such as in caramels, by reducing their tendency to stick to cutting knives, wrappers, and teeth.

Emulsifiers are also classified according to their solubility, being hydrophilic (water-loving) or lipophilic (oil-loving). The HLB, or hydrophilic/lipophilic balance, is a measure of the emulsifier's affinity to oil or water. The HLB range is 0 to 20, where 0 indicates completely lipophilic and 20 indicates completely hydrophilic. The behavior of emulsifiers in water according to the HLB range is shown in Table 6.

Table 7 lists the FDA names of food-grade emulsifiers and the Title 21 Code of Federal Regulations reference number for each.

Fats and Oils

Fats and oils belong to a group of substances classified as lipids which consist of the higher fatty acids and compounds associated with them. Lipids are characterized by their solubility in fat solvents, their insolubil-

Table 6. BEHAVIOR OF EMULSIFIERS IN WATER
ACCORDING TO   HLB RANGE

Behavior When Added to Water	HLB Range
Insoluble	1–4
Poor solubility	3–6
Milky dispersion after vigorous agitation	6–8
Stable milky dispersion	8–10
Translucent-to-clear dispersion	10–13
Clear solution	13+

HLB = Hydrophilic/lipophilic balance.

Table 7. FDA NAMES OF FOOD-GRADE EMULSIFIERS
AND THE TITLE 21 CODE OF FEDERAL
REGULATIONS REFERENCES

Emulsifier	Reference No.
Acetylated monoglycerides	172.828
Calcium Stearoyl-2-lactylate	172.844
Diacetyl tartaric acid esters of mono- and diglycerides	182.4101
Dioctyl sodium sulfosuccinate	172.810
Ethoxylated mono- and diglycerides	172.834
Fatty acids	172.860
Glyceryl-lacto esters of fatty acids	172.852
Hydroxylated lecithin	172.814
Lactylated fatty acid esters of glycerol and propylene glycol	172.850
Lactylic esters of fatty acids	172.848
Lecithin	184.1400
Mono- and diglycerides	182.4505
Monosodium phosphate derivatives of mono- and diglycerides	182.4521
Polyglycerol esters of fatty acids	172.854
Polyoxyethylene (20) sorbitan tristearate	172.838
Polyoxyethylene (20) sorbitan monostearate	172.836
Polyoxyethylene (20) sorbitan monooleate	172.840
Propylene glycol mono- and diesters of fats and fatty acids	172.856
Sodium lauryl sulfate	172.822
Sodium stearoyl-2-lactylate	172.846
Sodium stearyl fumarate	172.826
Sorbitan monostearate	172.842
Succinylated monoglycerides	172.830
Succistearin (stearoyl propylene glycol hydrogen succinate)	172.765

ity in water, and their greasy feel. Fats and oils are of similar chemical structure but differ physically in that at ordinary temperatures, fats are solids and oils are liquids. They are complex mixtures of predominantly mixed triglycerides, which are the compounds formed by combining one molecule of glycerol with three molecules of fatty acids. The fatty acids may be the same, two different fatty acids, or all different. Fatty acids are composed of a chain of carbon with hydrogen atoms, terminating in a carboxyl group. Fatty acids contain carbon chain lengths ranging from 4 to 24 and are identified according to the number of carbon atoms and whether they are saturated or unsaturated. Saturated fatty acids contain only single-bond carbon linkages and cannot accept additional hydrogen; unsaturated fatty acids have one or more double bonds and thus fewer hydrogen atoms and can accept hydrogen. Mono-unsaturated indicates that hydrogen can be accepted at one double-bond site; poly-unsaturated indicates that hydrogen can be accepted at more than one double-bond site. The most highly unsaturated fats are oils, while fats of low unsaturation tend to be solids at room temperature. Hydrogenation (chemical addition of hydrogen to the double bond of unsaturated fatty acids) of a fat makes it firmer and more plastic, raises the melting point, and slows the development of rancidity by reducing the rate of reaction with oxygen. These fats are termed hydrogenated or partially hydrogenated oils. Fats and oils are composed of varying percentages of fatty acids which account for their respective properties. Some useful properties to consider in determining a suitable fat or oil include:

1. Iodine value—an expression of the degree of unsaturation, which can serve as a guide in evaluating fat stability;
2. Melting point—the temperature at which a solid changes to a liquid when heated; and
3. Solid fat index—a number indicating the proportion of solid to liquid present in the fat at a given temperature, which will reflect the consistency.

Fats and oils are used by themselves or as components of a food. Some examples of nomenclature are:

1. Shortenings—usually solid fats instead of oils used in baked goods to impart tenderness, soft crumb, etc.;
2. Spreads—usually butter or margarine, which contains 80 percent or more fat;
3. Salad oils—oils which include olive, corn, cottonseed, soybean, sunflower; cooking fat—any edible fat or oil;
4. Frying fat—a bland-flavored fat or oil of high smoking temperature to allow for heating to 400°F without smoking; and

**Table 8. FATS AND OILS AND THEIR PRINCIPAL
COMPONENT FATTY ACIDS**

Fats and Oils	Approximate Percentage of the Principal Fatty Acids
Butter Fat	32 Oleic, 25 Palmitic
Cocoa Butter	37 Oleic, 35 Stearic, 25 Palmitic
Coconut	48 Lauric, 18 Myristic
Corn	58 Linoleic, 26 Oleic
Cottonseed	54 Linoleic, 24 Palmitic
Lard	46 Oleic, 23 Palmitic
Olive	68 Oleic, 14 Palmitic, 12 Linoleic
Palm	47 Palmitic, 38 Oleic
Palm Kernel	50 Lauric, 16 Myristic
Peanut	46 Oleic, 31 Linoleic
Rapeseed	31 Oleic, 23 Erucic, 19 Linoleic
Safflower	75 Linoleic
Sesame	43 Linoleic, 42 Oleic
Sorghum	52 Linoleic, 31 Oleic
Soybean	50 Linoleic, 25 Oleic
Sunflower	68 Linoleic, 20 Oleic
Tallow, beef	44 Oleic, 35 Palmitic
Tallow, mutton	43 Oleic, 30 Stearic

5. Confectionary fat—a fat that is hard at room temperature and soft at body temperature, such as hydrogenated coconut oil or cacao butter.

Table 8 lists fats and oils with their principal component fatty acids.

Flavors

Flavors are classified into the major groups of spices, natural flavors, and artificial flavors. A spice is an aromatic vegetable substance in a whole, broken, or ground form which is used as a seasoning. Natural flavors are flavor constituents derived from plant or animal sources. Artificial flavors are flavorings containing all or some portion of nonnatural materials.

Materials that can be used for flavorings can be grouped as follows: spices and herbs; essential oils and their extracts; fruits and fruit juices; and aliphatic, aromatic, and terpene compounds. Spices and herbs consist of dried plant products that exhibit flavor and aroma. They are derived from true aromatic vegetable substances from which the volatile and flavoring principles have not been removed. Essential oils and their extracts are odorous oils obtained from plant material and have the major odor that is characteristic of that material. Most have poor water solubility and most contain terpenes (hydrocarbons of formula $C_{10}H_{16}$ and their oxygenated derivatives $C_{10}H_{16}O$ or $C_{10}H_{18}O$) which contribute to the poor water solubility as well as possibly contributing to the off-flavor. Examples are essential oils of bitter almond, anise, and clove. Terpeneless oils are extensions of concentrated essential oils in which the unwanted terpenes are removed. These oils are usually more concentrated and of increased stability and water solubility. Common oils in the terpeneless form are citrus oil, spearmint, and peppermint. Fruit and fruit juices are natural flavorings obtained from fruits. Whole, crushed, or pureed fruit may be used, but, more commonly, the juice or concentrate is used. Fruit extracts are made by extraction with a water-alcohol mixture. Aliphatic, aromatic, and terpene compounds refer to synthetic chemicals and isolates from natural materials. This classification encompasses the largest group of flavoring materials.

Flavors can be quite complex and the number of available flavors is extensive. Flavor is that property of a substance that causes a sensation of taste. Four basic tastes are perceived by taste buds on the tongue: sweet, salty, sour, and bitter. The flavors used are natural, artificial, or combinations and exist in liquid or dry form. General flavor types available include fruit, dairy, meat, vegetable, beverage, and liquor.

Flour

Flour, also referred to as white flour, wheat flour, and plain flour, is the food prepared by grinding and bolting cleaned wheat other than durum wheat and red durum wheat (Code of Federal Regulations). Flour from other sources is available, identified according to its grain source. The properties of wheat flour vary according to the type of wheat, milling procedure, and treatment applied after milling.

Flour milling involves the separation of the endosperm, which is about 83 percent of the kernel, from other parts of the kernel, that is, bran and germ. The processing involves tempering, grinding, and sifting the large chunks of endosperm or "middlings" to yield the flour which, in the United States, will represent about 72 percent of the wheat kernel. By processing, size classification is achieved as flour streams. The streams include:

1. Straight flour—all the flour that can be milled from a wheat blend, or 72 percent of the wheat kernel which equates to 100 percent separation;
2. Long patent flour—90 to 95 percent separation;
3. Medium patent flour—80 to 90 percent separation;

4. Short patent flour—70 to 80 percent separation;
5. Short family or first patent flour—60 to 70 percent separation;
6. Extra short or fancy patent flour—40 to 60 percent separation; and
7. Clears—portion of straight flour remaining after removal of patent streams.

Flour properties depend upon the type of wheat, which is classified as hard or soft. Hard wheats are high in protein and the resulting flours have a high protein content that forms a tenacious, elastic gluten with good gas-retaining properties and high water absorption capacity which makes it suitable for yeast-leavened bread. Soft wheats are low in protein and the resulting flour has poor gas-retaining properties and low water absorption capacity which makes it suitable for chemically leavened cakes and pastries.

The protein content of the flour is important because it forms the protein complex termed *gluten* when water and flour are kneaded together. The gluten formed accounts for the mixing and dough-handling characteristics as well as the formation of the framework of the baked product.

Different flours are used for different purposes. Some of these include:

1. Bread flour, which generally contains in excess of 10.5 percent protein and is obtained from straight or long patent flours, has high absorption and good mixing tolerance;
2. Cake flour, which generally contains less than 10 percent protein and is generally short patent flours, is low in absorption, and has short mixing time and tolerance;
3. All-purpose (family) flour, which is intermediate between bread and cake flour;
4. Pastry flour, which is obtained from soft wheat and can be straight or clear flour grades because color is not an essential requirement; and
5. Cracker flour, which generally contains 9 to 10.5 percent protein obtained from long patent or straight flours, is of low absorption, and has short mixing requirements.

Flours can be modified by various treatments to alter characteristics such as color, nutritional value, and baking qualities. Some of these modified flours include:

1. Enriched flour—has been enriched by the inclusion of vitamins and minerals;

2. Bromated flour — potassium bromate has been added for improvement of baking qualities;
3. Phosphated flour — monocalcium phosphate has been added for improvement of baking qualities;
4. Bleached flour — flour in which the yellow carotenoid pigment has been converted to a nearly colorless product.

Other specific grain flours are obtained, with the term *flour* referring to that degree of grinding and sifting which results in a fine, powdery substance. The grain flours in Table 9 are designated according to the grain from which they are obtained and include corn, rye, and durum flours.

Gums

Gums, or hydrocolloids, are polysaccharides that function as water-control agents in increasing viscosity (resistance to flow) or forming gels. Gums are classified by source according to the following principal groupings: Plant exudates, which include arabic, tragacanth, karaya, ghatti; seaweed extracts, which include agar, alginates, carrageenan, furcellaran; plant seed gums, which include guar, locust bean, tamarind, psyllium, quince; plant extracts, which include pectin and arabinogalactan; fermentation gums, which include xanthan gum and dextran; and

Table 9. CEREAL FLOURS LISTED IN THE CODE OF FEDERAL REGULATIONS PART 137

Flour	Enriched corn grits
Bromated flour	Quick grits
Enriched bromated flour	Yellow grits
Enriched flour	White corn meal
Instantized flours	Bolted white corn meal
Phosphated flour	Enriched corn meals
Self-rising flour	Degerminated white corn meal
Enriched self-rising flour	Self-rising white corn meal
Cracked wheat	Yellow corn meal
Crushed wheat	Bolted yellow corn meal
Whole wheat flour	Degerminated yellow corn meal
Bromated whole wheat flour	Self-rising yellow corn meal
White corn flour	Farina
Yellow corn flour	Enriched farina
Durum flour	Semolina
Whole durum flour	Enriched rice
Corn grits	

cellulose derivatives, which include carboxymethyl cellulose, hydroxy-propylmethyl cellulose, microcrystalline cellulose. Gum derivatives includes propylene glycol alginate and low-methoxyl pectin. While starches and gelatin function as water-control agents, they are not included in this grouping.

The selection of a gum is based on the desired function and food application. By thickening or gelling the water, gums perform numerous roles such as stabilizers, film formers, binders, suspending agents, whipping agents, coating agents, and crystallization inhibitors. The gums perform these functions by themselves or in combination with other gums. Food properties considered in selecting a gum include pH, shelf stability, ingredient compatibility, texture, processing requirements, and ultimate consumer method of preparation. The differentiating properties of gums include viscosity, compatibility, pH stability, gel-forming capabilities, temperature stability, flow properties, and solubility. Within the same family of gums, there may be differences relative to salt type which will have an effect on its functional characteristics, that is, solubility, dispersibility, gel-forming capabilities, flow properties, and stability. Tables 10 and 11 list, respectively, comparative and relative properties of gums.

Preservatives

Preservatives are antimicrobial agents. The preservatives most widely used are the benzoates (sodium benzoate), sorbates (sorbic acid and potassium sorbate), and the propionates (sodium or calcium propionate), which are organic acids or their salts (see Table 12). The activity of preservatives is due to the undissociated form of the molecule and thus pH is a major factor in their effectiveness. Increasing the acidity of foods is a method of controlling the growth of microorganisms. The survival and proliferation of microorganisms depends in part upon the pH of the food. Foods with a pH below 4.5 to 4.6 are considered acidic, and many bacteria will not proliferate in acidic foods. Acidulants are used to reduce the pH and thus provide a means of controlling microorganism growth. Acidulants used include acetic acid, adipic acid, citric acid, fumaric acid, lactic acid, and phosphoric acid. Greater effectiveness is achieved in acidic systems. Sorbic acid and potassium sorbate have the best activity up to pH 6.5, calcium and sodium propionate up to pH 5.0, and sodium benzoate up to pH 4.5.

Sorbic acid and potassium sorbate are effective against yeast and mold inhibition with little activity against bacteria. Common uses are in cheese, sausage, and baked goods not including yeast-raised goods.

Table 10. COMPARATIVE PROPERTIES OF GUMS

Gums	Cold Water Solubility	Hot Water Solubility	Gel Former	Acid Stability
Agar	No	Yes	Yes	Between pH 4.5–9.0
Alginate, sodium	Yes	Yes	Yes	Gels at pH 3.5 depending on calcium content
Arabic (Acacia)	Yes	Yes	No	pH 4–10
Carboxymethyl cellulose	Yes	Yes	No	Best between pH 7–9, below pH 5 get reduction in viscosity
Carrageenan	No, except lambda & sodium salts	Yes	Yes, except lambda	Solution undergoes hydrolysis at acid pH (3.5); gel is stable
Furcellaren	No	Yes	Yes	Heating below pH 5 causes hydrolysis and gel degradation
Gelatin	No— swells	Yes	Yes	Stable, gradual decline in gel strength with acidification
Ghatti	Yes	Yes	No	Opt. viscosity at pH 8, drops on both sides

Guar	Yes	Yes	No	Between pH 3.5–10.5, gradual decline with acidification
Hydroxypropyl cellulose	Yes	No, insoluble above 45°C	No	Between pH 3–10, opt. pH 6–8
Hydroxypropyl methylcellulose	Yes	No	Yes, at elevated temp. depending on type	pH 3–11
Karaya	Yes, swells	Some	No	Viscosity decreased by acids or electrolytes
Locust bean gum	Swells, requires heat	Yes	No	Between pH 5–8, at higher or lower values get considerable variation
Low-methoxyl pectin	Yes, depends on methoxyl content	Yes	Yes	Form gels between pH 2.5–6.5 depending on system
Methylcellulose	Yes	No	Yes, at elevated temp	pH 3–11
Microcrystalline cellulose	Insoluble, dispersible	Insoluble, dispersible	No	Insoluble, resistant
Pectin	Yes	Yes	Yes	Gel below pH 3.6
Psyllium	Yes	Yes	Yes, at high conc.	Between pH 2–10
Quince	Yes	Yes	No	Between pH 4–10
Tragacanth	Yes, swells	Yes, swells	No	Between pH 4–6
Xanthan gum	Yes	Yes	No	Between pH 2–12

Table 11. RELATIVE PROPERTIES OF GELLING GUMS

Property	Agar	Carregeenan	Furcellaran	Gelatin	LM Pectin	Pectin	Sodium Alginate	Xanthan Gum/Locust Bean Gum
Solubility	>90°C	Kappa 50–60°C Iota 50–60°C Lambda room temp.	70–80°C	70°C	Room temp.	Room temp.	Room temp.	70°C
Gelling Temperature	Set 32–39°C Melt >85–95°C	Kappa, Iota—set & melt temp. are 10°C apart, vary with solute; Lambda: nongelling	Set—40°C, Melt—50°C, vary with solute	Set 20°C Melt 30°C	At some fixed temp. depending on system	Set 50–99°C Melt 70–100°C+ depends on type of pectin, pH, soluble solids, etc.	Room temp.	49–55°C

176

Mechanism	Cooling	K, Ca ions	Cooling	Cooling	Ca ions	Cooling; pH/soluble solids	Ca ions	Cooling
Reversibility	Thermo-reversible	Thermo-reversible	Thermo-reversible	Thermo-reversible	Thermo-reversible	Nonreversible; will melt but not reset	Non-reversible	Thermo-reversible
Transparency	Turbid	Kappa—transparent in pot. form. Iota—transparent in both forms	Pot. form—transparent. Ca form—turbid	Transparent	Turbid, transparent	Clear	Clear	Turbid
Texture	Brittle	Kappa—brittle Iota—elastic	Brittle/elastic	Elastic	Brittle	Brittle/elastic	Elastic	Elastic
Syneresis	Yes	Kappa—yes Iota—no	Yes	No	Yes	Some	Some	No

Table 12. PRESERVATIVES

Preservative Category	Activity	Use Level
Benzoates (sodium benzoate)	Yeasts, molds & bacteria, but usually not recommended for bacterial control because of restricted use level & lower activity at higher pH; best activity at pH 2.5–4.0	0.01–0.10%
Sorbates (sorbic acid, potassium sorbate)	Yeasts & molds, least activity against bacteria & on a selective basis; best activity up to pH 6.5	0.03–0.10%
Parabens	Yeasts & molds, less active against bacteria, especially gram-negative; effective up to pH 8.0	0.10% range
Propionates (propionic acid, Ca, Na propionate)	Molds, slight antibacterial action except against "rope"; effective up to pH 6.0	0.20–0.50%

Sorbic acid has low solubility in water which increases with increasing temperature, while potassium sorbate is readily soluble in water. Potassium sorbate has the same antimycotic properties as sorbic acid and on an equivalent weight basis has 74 percent of the activity of sorbic acid. Thus, higher concentrations are required to obtain the same yeast and mold-inhibiting effects (4 parts potassium sorbate equal 3 parts sorbic acid).

Calcium and sodium propionate are effective against molds and have slight antibacterial action and little action on yeasts. Application areas include baked goods and processed cheese. Because they have little action against yeasts, they can be used in yeast-baked goods and are the most common preservatives in baked goods.

Sodium benzoate is effective against yeasts and slightly effective against bacteria and molds. The most effective range is pH 2.5 to 4.0 with a maximum pH of 4.5. It is used in acidulated beverages, jams, jellies, and relishes.

Parabens, which are esters of para-hydroxybenzoic acid, are related to benzoic acid but are effective over a wider pH range. They are active against yeasts and molds and are used in baked goods and beverages.

Spices

Spices consist of dried plant products that exhibit flavor and aroma. Spices are obtained from vegetable substances from which none of the volatile or other flavoring substances have been removed. Spices are grouped into (a) tropical spices such as pepper and cloves; (b) herbs, such as sage and rosemary; (c) spicy seeds such as mustard and anise; and (d) dehydrated aromatic vegetables such as onion and garlic.

Spices in the ground form have an increased surface area and consequently the oil glands are ruptured, causing the evaporation of the essential oil and loss of aroma. Spice flavor is also obtained by use of an extract or essential oil which carries the spice aroma in a concentrated form. These oils are volatile and as such do not contain the nonvolatile constituents. The compound containing both the essential oil and the nonvolatile constituents is commercially known as oleoresins, which contain all the odorous and flavor principles of the spice. Oleoresins offer flavor uniformity, stability, freedom from bacteria, and flavor concentration.

Spices are used predominantly in prepared meats, luncheon meats, sauces, salads, soups, and dressings. Other important users are bakers, pickle packers, condiment manufacturers, and the canning industry.

A list of spices and other natural seasonings and flavorings is found in Title 21, Section 182.10, of the Code of Federal Regulations.

Starch

Starch, consisting of repeating glucose units, is separated into the polysaccharides amylose and amylopectin. Amylose consists of straight chains containing 200 to 2100 glucose units, while amylopectin consists of branched chains containing 20 to 25 glucose units each. A visible difference is that amylose is more soluble and less viscous than amylopectin and facilitates gel formation. Starches vary in their amylose content.

Waxy starches, so termed because the cut endosperm resembles hard, opaque wax, contain mostly amylopectin, while ordinary cornstarch consists of about 24 percent amylose and 76 percent amylopectin. The waxy starches form thick, clear pastes but gel only at high concentrations such as 30 percent, while 4 to 5 percent cornstarch will form a gel. The paste viscosity of the waxy maize starch remains the same hot or cold. High-amylose starches contain 50 to 70 percent amylose and have unique properties for functioning as film formers, oxygen and fat barriers, quick-setting stable gels, and binders.

Starch is not soluble in cold water, but forms a suspension. Upon heating the suspension to the gelatinization temperature (60 to 70°C), the starch granules suddenly swell, the opaque suspension slowly becomes translucent upon continued heating, and the viscosity increases to the thickness of a boiled starch paste. Most starches require heating to 90°C in order to obtain a firm gel upon cooling, when the viscosity increases and may form a gel depending on the type of starch. During storage of a starch paste or gel, the molecules become less soluble and tend to aggregate and partially crystallize; the change is termed *retrogradation*, which is the opposite of gelatinization. The starch gel shrinks and some of the liquid separates from the gel. Retrogradation does not occur in waxy starches because they do not contain amylose.

Starches can be modified (they are then termed *modified starches*) by chemical modification or cross-linking, to provide desired properties not found in natural starch. By the production of cross-links or bridges from one starch molecule to another, starch can be made more resistant to hydrolysis, thus preventing the loss of viscosity. The resistance of starch to shear or mixing is directly proportional to the degree of cross-linking. Other properties obtained include viscosity control, freeze-thaw stability, heat resistance, and acid resistance. These starches find application in sauces, gravies, pie filling, frozen foods, and other products where specific properties are required.

Pregelantinized starch is a starch processed to swell to some degree in cold water unlike regular starch, which requires heating. The most common method involves heating a starch paste to its gelatinization temperature, drying on a drum dryer, and grinding the dried starch to a powder. Upon reconstitution with water, the pregelatinized starch has less thickening power and tendency to gel than pastes of the parent starch. This starch is used in applications requiring more rapid hydration or room temperature preparation, such as instant desserts, puddings, and soups. Based on abundance and cost, cornstarch is the most commonly used. Other available starches include grain sorghum, rice, wheat, potato, tapioca, arrowroot, and waxy varieties.

Sweeteners

Sweeteners can be classified as natural or artificial. The natural sweeteners are carbohydrates consisting of molecules of carbon, hydrogen, and oxygen. The simplest form of carbohydrate is the monosaccharide or simple sugar and includes glucose (dextrose), fructose (levulose), and galactose, which are six-carbon (hexose) sugars.

The combination of two monosaccharides forms a disaccharide sugar, which can also be formed by the breakdown of longer chain carbohydrates termed *polysaccharides*. The following combinations of monosaccharides form the respective disaccharides: glucose + fructose = sucrose; glucose + galactose = lactose; glucose + glucose = maltose.

A trisaccharide consists of three monosaccharides, such as raffinose which consists of galactose, glucose, and fructose. A tetrasaccharide such as stachyose consists of four monosaccharides. These more complicated sugars are not digestible so they are not used as sweeteners. A polysaccharide is a longer-chained carbohydrate which exists in digestible and nondigestible forms. The digestible forms are starch, a polymer of glucose units from which corn sweeteners are obtained; and glycogen, a polymer of glucose which is the carbohydrate reserve of animals. The nondigestible form includes cellulose, lignin, and gums such as pectin and algin. Polyhydric alcohols (polyols) in foods consist of glycerine, sorbitol, mannitol, and propylene glycol. All except propylene glycol exhibit a sweetness that is less than the sweetness of sugar. When used at low levels, the taste of the polyols is of minor consequence; but when used as a major ingredient, such as in sugarless chewing gum, the polyol is the major source of sweetness.

Xylitol, a natural sugar substitute, is obtained from xylose which is derived from the acid hydrolysis of xylan, obtained from birch trees.

Artificial sweeteners permitted in food are saccharin and aspartame.

Table 13 lists the sweeteners and their relative sweetnesses.

Table 13. SWEETNESS RELATIVE TO SUCROSE

Sweetener	Sweetness Level*
Saccharin	300–400
Aspartame	200
Fructose	150
Sucrose	100
Xylitol	100
Dextrose	70
Sorbitol	55
Mannitol	50
Galactose	32
Lactose	20

*As percent (sucrose = 100).

Table 14. VITAMIN FUNCTIONS AND SOURCES

Vitamins	Function	Sources
Fat Soluble		
Vitamin B$_1$ (retinol)	Necessary for cell growth, healthy skin; prevents night blindness	Green & yellow fruits & vegetables; eggs, butter, cheese
Vitamin D (calciferol)	Necessary for bone & teeth growth; deficiency causes rickets	Fish, liver, oil, vit. D milk, sunshine
Vitamin E (tocopherol)	Functions as antioxidant, preventing the oxidation of unsaturated fatty acids & protecting vitamins such as vit. A	Legumes, meat, eggs, whole grains
Vitamin K	Essential for blood clotting	Green leafy vegetables, liver, soybeans
Water Soluble		
Vitamin B$_1$ (thiamine)	Necessary for growth, fertility, lactation; deficiency causes beriberi	Pork, fish, cereal, beans, peas
Vitamin B$_2$ (riboflavin)	Necessary for growth; acts as coenzyme	Milk, cheese, eggs, poultry
Vitamin B$_6$ (pyridoxine hydrochloride)	Functions as coenzyme; involved in utilization of protein	Meat, corn, lima beans
Vitamin B$_{12}$ (cobalamin)	Necessary for normal functioning of cells	Meat, liver, dry milk
Niacin	Necessary for healthy cells & tissues; prevents pellagra	Meat, liver, enriched bread
Pantothenic acid	Necessary for several bodily functions	Whole grain cereal, meat, fish
Biotin	Necessary in metabolism	Peanuts, beans, eggs, meat
Folic acid	Necessary in metabolism; helps manufacture red blood cells	Leafy green vegetables, yeast, liver
Vitamin C (ascorbic acid)	Essential for healthy bones & teeth; contributes to resistance to infection; deficiency causes scurvy	Fresh fruit & vegetables

Vitamins

Vitamins are organic compounds that are essential for normal body growth and maintenance. They are classified into two groups: fat-soluble vitamins—vitamins A, D, E, and K; and water-soluble vitamins—vitamins B and C. Vitamins are measured in very low concentrations, such as 1 to 100 mg. Through biochemical action, they perform various functions in such processes as cell growth, normal digestion, manufacture of red blood cells, and absorption of calcium and phosphorus. Inadequate vitamin intake can be the result of food deficiency, increased vitamin requirements, and increased vitamin loss. The vitamins of determined importance include: A (retinol), B_1 (thiamine), B_2 (riboflavin), B_5 (pantothenic acid), B_6 (pyridoxine), B_{12} (cyanocobalamin), C (ascorbic acid), D (calciferol), E (tocopherol), K, niacin, folic acid, and biotin.

Table 14 describes the functions and provides sources for fat- and water-soluble vitamins.

Substances for Use in Foods

Listed Under Title 21 of the
Code of Federal Regulations

Part 73. Listing of Color Additives Exempt From Certification

These substances

may be safely used as diluents in color additive mixtures for food use exempt from certification, subject to the condition that each straight color in the mixture has been exempted from certification. If not exempted, the color is from a batch that has previously been certified and has not changed in composition since certification.

Subpart A — Foods
- 73.1 Diluents in color additive mixtures for food use exempt from certification
- 73.30 Annatto extract
- 73.40 Dehydrated beets (beet powder)
- 73.50 Ultramarine blue
- 73.75 Canthaxanthin
- 73.85 Caramel
- 73.90 β-Apo-8′-carotenal
- 73.95 β-Carotene
- 73.100 Cochineal extract; carmine
- 73.140 Toasted partially defatted cooked cottonseed flour
- 73.160 Ferrous gluconate
- 73.169 Grape color extract
- 73.170 Grape skin extract (enocianina)

73.200   Synthetic iron oxide
73.250   Fruit juice
73.260   Vegetable juice
73.275   Dried algae meal
73.295   Tagetes (Aztec marigold) meal and extract
73.300   Carrot oil
73.315   Corn endosperm oil
73.340   Paprika
73.345   Paprika oleoresin
73.450   Riboflavin
73.500   Saffron
73.575   Titanium dioxide
73.600   Turmeric
73.614   Turmeric oleoresin

Part 74. Listing of Color Additives
Subject to Certification

These color additives may be safely used for coloring foods according to specified uses and restrictions.

Subpart A — Foods
74.101   FD&C Blue #1
74.102   FD&C Blue #2
74.203   FD&C Green #3
74.250   Orange B
74.302   Citrus Red #2
74.303   FD&C Red #3
74.340   FD&C Red #40
74.705   FD&C Yellow #5
74.706   FD&C Yellow #6

Part 172. Food Additives Permitted
for Direct Addition to Food
for Human Consumption

The listed food additives are permitted for direct addition to food for human consumption. Their usage is under conditions of good manufacturing practice.

Subpart A — General Provisions
172.5   General provisions for direct food additives

Subpart B – Food Preservatives

172.105   Anoxomer
172.110   BHA
172.115   BHT
172.120   Calcium disodium EDTA
172.130   Dehydroacetic acid
172.135   Disodium EDTA
172.140   Ethoxyquin
172.145   Heptylparaben
172.150   4-Hydroxymethyl-2,6-di-*tert*-butyl-phenol
172.155   Natamycin (pimaricin)
172.160   Potassium nitrate
172.165   Quaternary ammonium chloride combination
172.170   Sodium nitrate
172.175   Sodium nitrite
172.177   Sodium nitrite used in processing smoked chub
172.180   Stannous chloride
172.185   TBHQ
172.190   THBP

Subpart C — Coatings, Films, and Related Substances

172.210   Coatings on fresh citrus fruit
172.215   Coumarone-indene resin
172.225   Methyl esters of fatty acids produced from edible fats and oils
172.230   Microcapsules for flavoring substances
172.235   Morpholine
172.250   Petroleum naphtha
172.255   Polyacrylamide
172.260   Oxidized polyethylene
172.275   Synthetic paraffin and succinic derivatives
172.280   Terpene resin

Subpart D — Special Dietary and Nutritional Additives

172.310   Aluminum nicotinate
172.315   Nicotinamide-ascorbic acid complex
172.320   Amino acids
172.325   Bakers yeast protein
172.330   Calcium pantothenate, calcium chloride double salt
172.335   D-Pantothenamide
172.340   Fish protein isolate

172.345 Folic acid (folacin)
172.350 Fumaric acid and salts of fumaric acids
172.365 Kelp
172.370 Iron-choline citrate complex
172.372 N-Acetyl-L-methionine
172.375 Potassium iodide
172.385 Whole fish protein concentrate
172.395 Xylitol
172.399 Zinc methionine sulfate

Subpart E—Anticaking Agents
172.410 Calcium silicate
172.430 Iron ammonium citrate
172.480 Silicon dioxide
172.490 Yellow prussiate of soda

Subpart F—Flavoring Agents and Related Substances
172.510 Natural flavoring substances and natural substances used in conjunction with flavors
172.515 Synthetic flavoring substances and adjuvants
172.520 Cocoa with dioctyl sodium sulfoccinate for manufacturing
172.530 Disodium guanylate
172.535 Disodium inosinate
172.560 Modified hop extract
172.575 Quinine
172.580 Safrole-free extract of sassafras
172.585 Sugar beet extract flavor base
172.590 Yeast-malt sprout extract

Subpart G—Gums, Chewing Gum Bases and Related Substances
172.610 Arabinogalactan
172.615 Chewing gum base
172.620 Carrageenan
172.623 Carrageenan with polysorbate 80
172.626 Salts of carrageenan
172.655 Furcelleran
172.660 Salts of furcelleran
172.695 Xanthan gum

Subpart H—Other Specific Usage Additives
172.710 Adjuvants for pesticide use dilutions

172.712 Dimethyl dialkyl ammonium chloride
172.715 Calcium lignosulfonate
172.720 Calcium lactobionate
172.725 Gibberellic acid and its potassium salt
172.730 Potassium bromate
172.735 Glycerol ester of wood rosin
172.755 Stearyl monoglyceridyl citrate
172.765 Succistearin (stearoyl propylene glycol hydrogen succinate)
172.770 Ethylene oxide polymer
172.775 Methacrylic acid-divinylbenzene copolymer

Subpart I—Multipurpose Additives

172.802 Acetone peroxides
172.804 Aspartame
172.806 Azodicarbonamide
172.808 Copolymer condensates of ethylene oxide and propylene oxide
172.810 Dioctyl sodium sulfosuccinate
172.812 Glycine
172.814 Hydroxylated lecithin
172.816 Methyl glucoside-coconut oil ester
172.818 Oxystearin
172.820 Polyethylene glycol (mean molecular weight 200–9,500)
172.822 Sodium lauryl sulfate
172.824 Sodium mono- and dimethyl naphthalene sulfonates
172.826 Sodium stearyl fumarate
172.828 Acetylated monoglycerides
172.830 Succinylated monoglycerides
172.832 Monoglyceride citrate
172.834 Ethoxylated mono- and diglycerides
172.836 Polysorbate 60
172.838 Polysorbate 65
172.840 Polysorbate 80
172.841 Polydextrose
172.842 Sorbitan monostearate
172.844 Calcium stearoyl-2-lactylate
172.846 Sodium stearoyl-2-lactylate
172.848 Lactylic esters of fatty acids
172.850 Lactylated fatty acid esters of glycerol and propylene glycol
172.852 Glyceryl-lacto esters of fatty acids

172.854 Polyglycerol esters of fatty acids
172.856 Propylene glycol mono- and diesters of fats and fatty acids
172.858 Propylene glycol alginate
172.859 Sucrose fatty acid esters
172.860 Fatty acids
172.862 Oleic acid derived from tall oil fatty acids
172.863 Salts of fatty acids
172.864 Synthetic fatty alcohols
172.866 Synthetic glycerin produced by the hydrogenolysis of carbohydrates
172.868 Ethyl cellulose
172.870 Hydroxypropyl cellulose
172.872 Methyl ethyl cellulose
172.874 Hydroxypropyl methylcellulose
172.876 Castor oil
172.878 White mineral oil
172.880 Petrolatum
172.882 Synthetic isoparaffinic petroleum hydrocarbons
172.884 Odorless light petroleum hydrocarbons
172.886 Petroleum wax
172.888 Synthetic petroleum wax
172.890 Rice bran wax
172.892 Food starch-modified
172.894 Modified cottonseed products intended for human consumption
172.896 Dried yeasts
172.898 Bakers yeast glycan

Part 182. Substances Generally Recognized as Safe

The substances listed are generally recognized as safe (GRAS) for their intended use and purpose, when used in accordance with good manufacturing practices. Good Manufacturing Practice includes (a) use of a quantity that does not exceed the amount reasonably required to accomplish its intended function; (b) the reduced use to the extent reasonably possible of the quantity of a substance that becomes a component of a food as a result of its use in the manufacturing, processing, or packaging of food but is not intended to accomplish any physical or technical effect; and (c) use of substances of appropriate food grade and prepared and handled as food ingredients.

Subpart A—General Provisions

See specific sections below:

§182.10 SPICES AND OTHER NATURAL SEASONINGS AND FLAVORINGS

Common Name	Botanical Name of Plant Source
Alfalfa herb and seed	*Medicago sativa* L.
All spice	*Pimenta officinalis* Lindl
Ambrette seed	*Hibiscus abelmoschus* L.
Angelica	*Angelica archangelica* L. or other spp. of *Angelica*
Angelica root	"
Angelica seed	"
Angostura (cusparia bark)	*Galipea officinalis* Hancock
Anise	*Pimpinella anisum* L.
Anise, star	*Illicium verum* Hook. f.
Balm (lemon balm)	*Melissa officinalis* L.
Basil, bush	*Ocimum minimum* L.
Basil, sweet	*Ocimum basilicum* L.
Bay	*Laurus nobilis* L.
Calendula	*Calendula officinalis* L.
Camomile (chamomile), English or Roman	*Anthemis nobilis* L.
Camomile (chamomile), German or Hungarian	*Matricaria chamomilla* L.
Capers	*Cappans spinosa* L.
Capsicum	*Capsicum frutescens* L. or *Capsicum annuum* L.
Caraway	*Carum carvi* L.
Caraway, black (black cumin)	*Nigella sativa* L.

§182.10 (*cont.*)

Common Name	Botanical Name of Plant Source
Cardamom (cardamon)	*Elettaria cardamomum* Maton
Cassia, Chinese	*Cinnamomum cassia* Blume
Cassia, Padang or Batavia	*Cinnamomum burmanni* Blume
Cassia, Saigon	*Cinnamomum loureirii* Nees
Cayenne pepper	*Capsicum frutescens* L. or *Capsicum annuum* L.
Celery seed	*Apium graveolens* L.
Chervil	*Anthriscus cerefolium* (L). Hoffm.
Chives	*Allium schoenoprasum* L.
Cinnamon, Ceylon	*Cinnamomum zeylanicum* Nees
Cinnamon, Chinese	*Cinnamomum cassia* Blume
Cinnamon, Saigon	*Cinnamomum loureirii* Nees
Clary (clary sage)	*Salvia sclarea* L.
Clover	*Trifolium* spp.
Coriander	*Coriandrum sativum* L.
Cumin (cummin)	*Cuminum cyminum* L.
Cumin, black (black caraway)	*Nigella sativa* L.
Elder flowers	*Sambucus canadensis* L.
Fennel, common	*Foeniculum vulgare* Mill
Fennel, sweet (finocchio, Florence fennel)	*Foeniculum vulgare* Mill. var. duice (DC.) Alex
Fenugreek	*Trigonella foenum-graecum* L.
Galanga (galangal)	*Alpinia officinarum* Hance
Geranium	*Pelargonium* spp.
Ginger	*Zingiber officinale* Rosc.
Grains of paradise	*Amomum melegueta* Rosc.
Horehound (hoarhound)	*Marrubium vulgare* L.
Horseradish	*Armoracia lapathifolia* Gilib
Hyssop	*Hyssopus officinalis* L.
Lavender	*Lavandula officinalis* Chaix
Linden flowers	*Tilia* spp.
Mace	*Myristica fragrans* Houtt
Marigold, pot	*Calendula officinalis* L.
Marjoram, pot	*Marjorana onites* (L.) Benth
Marjoram, sweet	*Marjorana hortensis* Moench
Mustard, black or brown	*Brassica nigra* (L.) Koch
Mustard, brown	*Brassica juncea* (L.) Coss
Mustard, white or yellow	*Brassica hirta* Moench
Nutmeg	*Myristica fragans* Houtt
Oregano (oreganum, Mexican oregano, Mexican sage, origan)	*Lippia* spp.
Paprika	*Capsicum annuum* L.
Parsley	*Petroselinum crispum* (Mill.) Mansf.
Pepper, black	*Piper nigrum* L.
Pepper, cayenne	*Capsicum frutescens* L. or *Capsicum annuum* L.
Pepper, red	"

§182.10 (*cont.*)

Common Name	Botanical Name of Plant Source
Pepper, white	*Piper nigrum* L.
Peppermint	*Mentha piperita* L.
Poppy seed	*Papayer somniferum* L.
Pot marigold	*Calendula officinalis* L.
Pot marjoram	*Marjorana onites* (L.) Benth
Rosemary	*Rosmarinus officinalis* L.
Saffron	*Crocus sativus* L.
Sage	*Salvia officinalis* L.
Sage, Greek	*Salvia trioba* L.
Savory, summer	*Satureia hortensis* L. (Satureja)
Savory, winter	*Satureia montana* L. (Satureja)
Sesame	*Sesamum indicum* L.
Spearmint	*Mentha spicata* L.
Star anise	*Illicium verum* Hook. f.
Tarragon	*Artemisia dracunculus* L.
Thyme	*Thymus vulgaris* L.
Thyme, wild or creeping	*Thymus serpyllum* L.
Turmeric	*Curcuma longa* L.
Vanilla	*Vanilla planifolia* Andr. or *Vanilla tahitensis* J.W. Moore
Zedoary	*Curcuma zedoaria* Rosc.

§182.20 ESSENTIAL OILS, OLEORESINS (SOLVENT-FREE), AND NATURAL EXTRACTIVES (INCLUDING DISTILLATES)

Common Name	Botanical Name of Plant Source
Alfalfa	*Medicago sativa* L.
Allspice	*Pimenta officinalis* Lindl
Almond, bitter (from prussic acid)	*Prunus amygdalus* Batsch, *Prunus armeniaca* L., or *Prunus persica* (L.) Batsch
Ambrette (seed)	*Hibiscus moschatus* Moench
Angelica root	*Angelica archangelica* L.
Angelica seed	"
Angelica stem	"
Angostura (cusparia bark)	*Galipea officinalis* Hancock
Anise	*Pimpinella anisum* L.
Asafetida	*Ferula assa-foetida* L. and related spp. of *Ferula*
Balm (lemon balm)	*Melissa officinalis* L.
Balsam of Peru	*Myroxylon pereirae* Klotzsch
Basil	*Ocimum basilicum* L.
Bay leaves	*Laurus nobilis* L.
Bay (myrcia oil)	*Pimenta racemosa* (Mill.) J.W. Moore

Common Name	Botanical Name of Plant Source
Bergamot (bergamot orange)	*Citrus aurantium* L. subsp. bergamia Wright et Arn
Bitter almond (free from prussic acid)	*Prunus amygdalus* Batsch, *Prunus armeniaca* L., or *Prunus persica* (L.) Batsch
Bois de rose	*Aniba rosaeodora* Ducke
Cacao	*Theobroma cacao* L.
Camomile (chamomile) flowers, Hungarian	*Matricaria chamomilla* L.
Camomile (chamomile) flowers, Roman or English	*Anthemis nobilis* L.
Cananga	*Cananga odorata* Hook. f. and Thoms
Capsicum	*Capsicum frutescens* L. and *Capsicum annuum* L.
Caraway	*Carum carvi* L.
Cardamom seed (cardamon)	*Elettaria cardamomum* Maton
Carob bean	*Ceratonia siliqua* L.
Carrot	*Daucus carota* L.
Cascarilla bark	*Croton eluteria* Benn
Cassia bark, Chinese	*Cinnamomum cassia* Blume
Cassia bark, Padang or Batavia	*Cinnamomum burmanni* Blume
Cassia bark, Saigon	*Cinnamomum loureirii* Nees
Celery seed	*Apium graveolens* L.
Cherry, wild, bark	*Prunus serotina* Ehrh.
Chervil	*Anthriscus cerefolium* (L.) Hoffm.
Chicory	*Cichorium intybus* L.
Cinnamon bark, Ceylon	*Cinnamomum zeylanicum* Nees
Cinnamon bark, Chinese	*Cinnamomum cassia* Blume
Cinnamon bark, Saigon	*Cinnamomum loureirii* Nees
Cinnamon leaf, Ceylon	*Cinnamomum zeylanicum* Nees
Cinnamon leaf, Chinese	*Cinnamomum cassia* Blume
Cinnamon leaf, Saigon	*Cinnamomum loureirii* Nees
Citronella	*Cymbopogon nardus* Rendle
Citrus peels	*Citrus* spp.
Clary (clary sage)	*Salvia sclarea* L.
Clover	*Trifolium* spp.
Coca (decocainized)	*Erythroxylum coca* Lam. and other spp. of *Erythroxylum*
Coffee	*Coffea* spp.
Cola nut	*Cola acuminata* Schott and Endl., and other spp. of *Cola*
Coriander	*Coriandrum sativum* L.
Cumin (cummin)	*Cuminum cyminum* L.
Curacao orange peel (orange, bitter peel)	*Citrus aurantium* L.
Cusparia bark	*Galipea officinalis* Hancock
Dandelion	*Taraxacum officinale* Weber and *T. laevigatum* DC.

§182.20 (*cont.*)

Common Name	Botanical Name of Plant Source
Dandelion root	"
Dog grass (quackgrass, triticum)	*Agropyron repens* (L.) Beauv.
Elder flowers	*Sambucus canadensis* L. and *S. nigra* l.
Estragole (esdragol, esdragon, tarragon)	*Artemisia dracunculus* L.
Estragon (tarragon)	"
Fennel, sweet	*Foeniculum vulgare* Mill
Fenugreek	*Trigonella foenum-graecum* L.
Galanga (galangal)	*Alpinia officinarum* Hance
Geranium	*Pelargonium* spp.
Geranium, East Indian	*Cymbopogon martini* Stapf.
Geranium, rose	*Pelargonium graveolens* L'Her.
Ginger	*Zingiber officinale* Rosc.
Grapefruit	*Citrus paradisi* Macf.
Guava	*Psidium* spp.
Hickory bark	*Carya* spp.
Horehound (hoarhound)	*Marrubium vulgare* L.
Hops	*Humulus lupulus* L.
Horsemint	*Monarda punctata* L.
Hyssop	*Hyssopus officinalis* L.
Immortelle	*Helichrysum augustifolium* DC.
Jasmine	*Jasminum officinale* L. and other spp. of *Jasminum*
Juniper (berries)	*Juniperus communis* L.
Kola nut	*Cola acuminata* Schott and Endl., and other spp.of *Cola*
Laurel berries	*Laurus nobilis* L.
Laurel leaves	*Laurus* spp.
Lavender	*Lavandula officinalis* Chaix
Lavender, spike	*Lavandula latifolia* Vill
Lavandin	Hybrids between *Lavandula officinalis* Chaix and *Lavandula latifolin* Vill
Lemon	*Citrus limon* (L.) Burm. f.
Lemon balm (see balm)	
Lemon grass	*Cymbopogon citratus* DC. and *Cymbopogon lexuosus* Stapf.
Lemon peel	*Citrus limon* (L.) Burm. f.
Lime	*Citrus aurantifolia* Swingle
Linden flowers	*Tilia* spp.
Locust bean	*Ceratonia siliqua* L.
Lupulin	*Humulus lupulus* L.
Mace	*Myristica fragans* Houtt
Mandarin	*Citrus reticulata* Blanco
Marjoram, sweet	*Marjorana hortensis* Moench
Mate	*Ilex paraguariensis* St. Hil.
Melissa (see balm)	

§182.20 (*cont.*)

Common Name	Botanical Name of Plant Source
Menthol	*Mentha* spp.
Menthyl acetate	"
Molasses (extract)	*Saccarum officinarium* L.
Mustard	*Brassica* spp.
Maringin	*Citrus paradisi* Macf.
Neroli, bigarade	*Citrus aurantium* L.
Nutmeg	*Myristica fragans* Houtt
Onion	*Allium cepa* L.
Orange, bitter, flowers	*Citrus aurantium* L.
Orange, bitter, peel	"
Orange leaf	*Citrus sinensis* (L.) Osbeck
Orange, sweet	"
Orange, sweet, flowers	"
Orange, sweet, peel	"
Origanum	*Origanum* spp.
Palmarosa	*Cymbopogon martini* Stapf.
Paprika	*Capsicum annuum* L.
Parsley	*Petroselinum crispum* (Mill.) Mansf.
Pepper, black	*Piper nigrum* L.
Pepper, white	"
Peppermint	*Mentha piperita* L.
Peruvian balsam	*Myroxylon pereirae* Klotzsch
Petitgrain	*Citrus aurantium* L.
Petitgrain lemon	*Citrus limon* (L.) Burm. f.
Petitgrain mandarin or tangerine	*Citrus reticulata* Blanco
Pimenta	*Pimenta officinalis* Lindl
Pimenta leaf	*Pimenta officinalis* Lindl
Pipsissewa leaves	*Chimaphila umbellata* Nutt
Pomegranate	*Punica granatum* L.
Prickly ash bark	*Xanthoxylum* (or *Zanthoxylum*) *Americanum* Mill. or *Xanthoxylum clava-herculis* L.
Rose absolute	*Rosa alba* L., *Rosa centifolia* L., *Rosa damascena* Mill., *Rosa gallica* L., and vars. of these spp.
Rose (otto of roses, attar of roses)	"
Rose buds	"
Rose flowers	"
Rose fruit (hips)	"
Rose geranium	*Pelargonium graveolens* L'Her.
Rose leaves	*Rosa* spp.
Rosemary	*Rosmarinus officinalis* L.
Saffron	*Crocus sativus* L.
Sage	*Salvia officinalis* L.
Sage, Greek	*Salvia triloba* L.
Sage, Spanish	*Salvia lavandulaefolia* Vahl
St. John's bread	*Ceratonia siliqua* L.
Savory, summer	*Satureia hortensis* L.

§182.20   (*cont.*)

Common Name	Botanical Name of Plant Source
Savory, winter	*Satureia montana* L.
Schinus molle	*Schinus molle* L.
Sloe berries (blackthorn berries)	*Prunus spinosa* L.
Spearmint	*Mentha spicata* L.
Spike lavender	*Lavandula latifolia* Vill.
Tamarind	*Tamarindus indica* L.
Tangerine	*Citrus reticulata* Blanco
Tarragon	*Artemisia dracunculus* L.
Tea	*Thea sinensis* L.
Thyme	*Thymus vulgaris* L. and *Thymus zygis* var. *gracilis* Boiss
Thyme, white	"
Thyme, wild or creeping	*Thymus serpyllum* L.
Triticum (see dog grass)	
Tuberose	*Polianthes tuberosa* L.
Turmeric	*Curcuma longa* L.
Vanilla	*Vanilla planifolia* Andr. or *Vanilla tahitensis* J.W. Moore
Violet flowers	*Viola odorata* L.
Violet leaves	"
Violet leaves absolute	"
Wild cherry bark	*Prunus serotina* Ehrh.
Ylang-ylang	*Cananga odorata* Hook. f. and Thoms
Zedoary bark	*Curcuma zedonaria* Rosc.

§182.40 NATURAL EXTRACTIVES (SOLVENT-FREE) USED IN CONJUNCTION WITH SPICES, SEASONINGS, AND FLAVORINGS

Common Name	Botanical Name of Plant Source
Apricot kernel (persic oil)	*Prunus armeniaca* L.
Peach kernel (persic oil)	*Prunus persica* Sieb. et Zucc.
Peanut stearine	*Arachis hypogaea* L.
Persic oil (see apricot kernel and peach kernel)	
Quince seed	*Cydonia oblonga* Miller

§182.50 CERTAIN OTHER SPICES, SEASONINGS, ESSENTIAL OILS, OLEORESINS, AND NATURAL EXTRACTS

Common Name	Derivation
Ambergris	*Physeter macrocephalus* L.
Castoreum	*Castor fiber* L. and *C. canadensis* Kuhl
Civet (zibeth, zibet, zibetum)	Civet cats, *Viverra civetta* Schreber and *Viverra zibetha* Schreber
Cognac oil, white and green	Ethyl oenanthate, so-called
Musk (Tonquin musk)	Musk deer, *Moschus moschiferus* L.

§182.60 SYNTHETIC FLAVORING SUBSTANCES AND ADJUVANTS

Acetaldehyde (ethanol)
Acetoin (acetyl methylcarbinol)
Anethole (parapropenyl anisole)
Benzaldehyde (benzoic aldehyde)
N-Butyric acid (butanoic acid)
D- or *L*-Carvone (carvol)
Cinnamaldehyde (cinnamic aldehyde)
Citral (2,6-dimethyloctadien-2,6-*al*-8, geranial, neral)
Decanal (*N*-decylaldehyde, capraldehyde, capric aldehyde, caprinaldehyde, aldehyde C-10)
Ethyl acetate
Ethyl butyrate

3-Methyl-3-phenyl glycidic acid ethyl ester (ethyl-menthyl-phenyl-glycidate, so-called strawberry aldehyde, C-16 aldehyde)
Ethyl vanillin
Geraniol (3,7-dimethyl-2,6 and 3,6-octadien-1-*ol*)
Geranyl acetate (geraniol acetate)
Glycerol (glyceryl) tributyrate (tributyrin, butyrin)
Limonene (*D*-, *L*-, and *DL*-)
Linalool (linalol, 3,7-dimethyl-1,6-octadien-3-*ol*)
Linalyl acetate (bergamol)
Methyl anthranilate (methyl-2-aminobenzoate)
Piperonal (3,4-methylenedioxy-benzaldehyde, heliotropin)
Vanillin

§182.70 SUBSTANCES MIGRATING FROM COTTON AND COTTON FABRICS USED IN DRY FOOD PACKAGING

Beef tallow
Carboxymethylcellulose
Coconut oil, refined
Cornstarch
Gelatin
Japan wax
Lard
Lard oil
Oleic acid
Peanut oil
Potato starch
Sodium acetate
Sodium chloride
Sodium silicate
Sodium tripolyphosphate
Soybean oil (hydrogenated)
Talc
Tallow (hydrogenated)
Tallow flakes
Tapioca starch
Tetrasodium pyrophosphate
Wheat starch
Zinc chloride

§182.90 SUBSTANCES MIGRATING TO FOOD FROM PAPER AND PAPERBOARD PRODUCTS

Alum (double sulfate of aluminum and ammonium potassium, or sodium)
Aluminum hydroxide
Aluminum oleate
Aluminum palmitate
Casein
Cellulose acetate
Cornstarch
Corn sugar (syrup)
Diatomaceous earth filler
Ethyl cellulose
Ethyl vanillin
Ferric sulfate
Ferrous sulfate
Glycerin
Invert sugar
Iron, reduced

§182.90 (*cont.*)

Mono- and diglycerides from glycerolysis of edible fats and oils
Oleic acid
Oxides of iron
Potassium sorbate
Silicon dioxides
Sodium aluminate
Sodium chloride
Sodium hexametaphosphate
Sodium hydrosulfite
Sodium phosphoaluminate
Sodium silicate
Sodium sorbate
Sodium tripolyphosphate
Sorbitol
Soy protein, isolated
Starch, acid-modified
Starch, pregelatinized
Starch, unmodified
Sucrose
Talc
Vanillin
Zinc hydrosulfite
Zinc sulfate

Subpart B—Multiple Purpose GRAS Food Substances

182.1033	Citric acid
182.1045	Glutamic acid
182.1047	Glutamic acid hydrochloride
182.1057	Hydrochloric acid
182.1073	Phosphoric acid
182.1087	Sodium acid pyrophosphate
182.1125	Aluminum sulfate
182.1127	Aluminum ammonium sulfate
182.1129	Aluminum potassium sulfate
182.1131	Aluminum sodium sulfate
182.1180	Caffeine
182.1195	Calcium citrate
182.1217	Calcium phosphate
182.1235	Caramel
182.1320	Glycerin
182.1324	Glyceryl monostearate
182.1480	Methylcellulose
182.1500	Monoammonium glutamate
182.1516	Monopotassium glutamate

182.1625   Potassium citrate
182.1711   Silica aerogel
182.1745   Sodium carboxymethylcellulose
182.1748   Sodium caseinate
182.1751   Sodium citrate
182.1778   Sodium phosphate
182.1781   Sodium aluminum phosphate
182.1810   Sodium tripolyphosphate
182.1866   High-fructose corn syrup
182.1901   Triacetin
182.1911   Triethyl citrate

Subpart C—Anticaking Agents
182.2122   Aluminum calcium silicate
182.2227   Calcium silicate
182.2437   Magnesium silicate
182.2727   Sodium aluminosilicate
182.2729   Sodium calcium aluminosilicate, hydrated
182.2906   Tricalcium silicate

Subpart D—Chemical Preservatives
182.3013   Ascorbic acid
182.3041   Erythorbic acid
182.3089   Sorbic acid
182.3109   Thiodipropionic acid
182.3149   Ascorbyl palmitate
182.3169   Butylated hydroxyanisole
182.3173   Butylated hydroxytoluene
182.3189   Calcium ascorbate
182.3225   Calcium sorbate
182.3280   Dilauryl thiodipropionate
182.3616   Potassium bisulfite
182.3637   Potassium metabisulfite
182.3640   Potassium sorbate
182.3731   Sodium ascorbate
182.3739   Sodium bisulfite
182.3766   Sodium metabisulfite
182.3795   Sodium sorbate
182.3798   Sodium sulfite
182.3862   Sulfur dioxide
182.3890   Tocopherols

Subpart E—Emulsifying Agents
182.4101   Diacetyl tartaric acid esters of mono- and diglycerides
of edible fats or oils, or edible fat-forming fatty acids

182.4505   Mono- and diglycerides of edible fats or oils, or edible fat-forming acids
182.4521   Monosodium phosphate derivatives of mono- and diglycerides of edible fats or oils, or edible fat-forming fatty acids

Subpart F—Dietary Supplements

182.5013   Ascorbic acid
182.5065   Linoleic acid
182.5159   Biotin
182.5191   Calcium carbonate
182.5195   Calcium citrate
182.5201   Calcium glycerophosphate
182.5210   Calcium oxide
182.5212   Calcium pantothenate
182.5217   Calcium phosphate
182.5223   Calcium pyrophosphate
182.5245   Carotene
182.5250   Choline bitartrate
182.5252   Choline chloride
182.5260   Copper gluconate
182.5301   Ferric phosphate
182.5304   Ferric pyrophosphate
182.5306   Ferric sodium pyrophosphate
182.5308   Ferrous gluconate
182.5311   Ferrous lactate
182.5315   Ferrous sulfate
182.5370   Inositol
182.5375   Iron reduced
182.5431   Magnesium oxide
182.5434   Magnesium phosphate
182.5443   Magnesium sulfate
182.5446   Manganese chloride
182.5449   Manganese citrate
182.5452   Manganese gluconate
182.5455   Manganese glycerophosphate
182.5461   Manganese sulfate
182.5464   Manganous oxide
182.5530   Niacin
182.5535   Niacinamide
182.5580   D-Pantothenyl alcohol
182.5628   Potassium glycerophosphate
182.5676   Pyridoxine hydrochloride

182.5695	Riboflavin
182.5697	Riboflavin-5-phosphate
182.5772	Sodium pantothenate
182.5778	Sodium phosphate
182.5875	Thiamine hydrochloride
182.5878	Thiamine mononitrate
182.5890	Tocopherols
182.5892	α-Tocopherol acetate
182.5930	Vitamin A
182.5933	Vitamin A acetate
182.5936	Vitamin A palmitate
182.5945	Vitamin B_{12}
182.5950	Vitamin D_2
182.5953	Vitamin D_3
182.5985	Zinc chloride
182.5988	Zinc gluconate
182.5991	Zinc oxide
188.5994	Zinc stearate
182.5997	Zinc sulfate

Subpart G — Sequestrants

182.6033	Citric acid
182.6085	Sodium acid phosphate
182.6195	Calcium citrate
182.6197	Calcium diacetate
182.6203	Calcium hexametaphosphate
182.6215	Monobasic calcium phosphate
182.6285	Dipotassium phosphate
182.6290	Disodium phosphate
182.6386	Isopropyl citrate
182.6511	Monoisopropyl citrate
182.6625	Potassium citrate
182.6751	Sodium citrate
182.6757	Sodium gluconate
182.6760	Sodium hexametaphosphate
182.6769	Sodium metaphosphate
182.6778	Sodium phosphate
182.6787	Sodium pyrophosphate
182.6789	Tetrasodium pyrophosphate
182.6810	Sodium tripolyphosphate
182.6851	Stearyl citrate

Subpart H — Stabilizers

182.7255	Chondrus extract

Subpart I—Nutrients

182.8013	Ascorbic acid
182.8159	Biotin
182.8195	Calcium citrate
182.8201	Calcium glycerophosphate
182.8217	Calcium phosphate
182.8223	Calcium pyrophosphate
182.8245	Carotene
182.8250	Choline bitartrate
182.8252	Choline chloride
182.8301	Ferric phosphate
182.8304	Ferric pyrophosphate
182.8306	Ferric sodium pyrophosphate
182.8308	Ferrous gluconate
182.8311	Ferrous lactate
182.8315	Ferrous sulfate
182.8375	Iron reduced
182.8455	Manganese glycerophosphate
182.8458	Manganese hypophosphite
182.8628	Potassium glycerophosphate
182.8778	Sodium phosphate
182.8890	Tocopherols
182.8892	α-Tocopherol acetate
182.8985	Zinc chloride
182.8988	Zinc gluconate
182.8991	Zinc oxide
182.8994	Zinc stearate
182.8997	Zinc sulfate

Part 184. Direct Food Substances Affirmed As Generally Recognized As Safe

The substances listed have been reviewed and determined by the Food and Drug Administration to be generally recognized as safe (GRAS) for the purposes and the other conditions prescribed. They shall be used in accordance with current good manufacturing practice.

Subpart A—General Provisions

184.1   Substances added directly to human food affirmed as generally recognized as safe (GRAS)

Subpart B—Listing of Specific Substances Affirmed as GRAS

184.1005   Acetic acid

184.1007	Aconitic acid
184.1009	Adipic acid
184.1011	Alginic acid
184.1021	Benzoic acid
184.1025	Caprylic acid
184.1027	Mixed carbohdrase and protease enzyme product
184.1061	Lactic acid
184.1065	Linoleic acid
184.1069	Malic acid
184.1077	Potassium acid tartrate
184.1081	Propionic acid
184.1090	Stearic acid
184.1091	Succinic acid
184.1095	Sulfuric acid
184.1097	Tannic acid
184.1099	Tartaric acid
184.1115	Agar-agar
184.1120	Brown algae
184.1121	Red algac
184.1133	Ammonium alginate
184.1135	Ammonium bicarbonate
184.1137	Ammonium carbonate
184.1138	Ammonium chloride
184.1139	Ammonium hydroxide
184.1141a	Ammonium phosphate, monobasic
184.1141b	Ammonium phosphate, dibasic
184.1143	Ammonium sulfate
184.1155	Bentonite
184.1157	Benzoyl peroxide
184.1165	*n*-Butane and isobutane
184.1185	Calcium acetate
184.1187	Calcium alginate
184.1191	Calcium carbonate
184.1193	Calcium chloride
184.1199	Calcium gluconate
184.1205	Calcium hydroxide
184.1206	Calcium iodate
184.1207	Calcium lactate
184.1210	Calcium oxide
184.1212	Calcium pantothenate
184.1221	Calcium propionate
184.1229	Calcium stearate
184.1230	Calcium sulfate

184.1240   Carbon dioxide
184.1257   Clove and its derivatives
184.1259   Cocoa butter substitute primarily from palm oil
184.1260   Copper gluconate
184.1261   Copper sulfate
184.1262   Corn silk and corn silk extract
184.1265   Cuprous iodide
184.1271   L-Cysteine
184.1272   L-Cysteine monohydrochloride
184.1277   Dextrin
184.1278   Diacetyl
184.1282   Dill and its derivatives
184.1293   Ethyl alcohol
184.1295   Ethyl formate
184.1317   Garlic and its derivatives
184.1318   Glucono delta-lactone
184.1321   Corn gluten
184.1322   Wheat gluten
184.1330   Acacia (gum arabic)
184.1333   Gum ghatti
184.1339   Guar gum
184.1343   Locust (carob) bean gum
184.1349   Karaya gum (sterculia gum)
184.1351   Gum tragacanth
184.1355   Helium
184.1366   Hydrogen peroxide
184.1370   Inositol
184.1372   Insoluble glucose isomerase enzyme preparations
184.1388   Lactase enzyme preparation from *Kluyveromyces lactis*
184.1400   Lecithin
184.1408   Licorice and licorice derivatives
184.1409   Ground limestone
184.1425   Magnesium carbonate
184.1426   Magnesium chloride
184.1428   Magnesium hydroxide
184.1431   Magnesium oxide
184.1434   Magnesium phosphate
184.1440   Magnesium stearate
184.1443   Magnesium sulfate
184.1444   Maltodextrin
184.1445   Malt syrup (malt extract)
184.1446   Manganese chloride

184.1449	Manganese citrate
184.1452	Manganese gluconate
184.1461	Manganese sulfate
184.1490	Methylparaben
184.1530	Niacin
184.1535	Niacinamide
184.1537	Nickel
184.1540	Nitrogen
184.1545	Nitrous oxide
184.1553	Peptones
184.1555	Rapeseed oil
184.1560	Ox bile extract
184.1563	Ozone
184.1585	Papain
184.1588	Pectins
184.1610	Potassium alginate
184.1613	Potassium bicarbonate
184.1619	Potassium carbonate
184.1622	Potassium chloride
184.1631	Potassium hydroxide
184.1634	Potassium iodide
184.1635	Potassium iodate
184.1643	Potassium sulfate
184.1655	Propane
184.1660	Propyl gallate
184.1666	Propylene glycol
184.1670	Propylparaben
184.1676	Pyridoxine hydrochloride
184.1685	Rennet (animal-derived)
184.1695	Riboflavin
184.1697	Riboflavin-5'-phosphate (sodium)
184.1698	Rue
184.1699	Oil of rue
184.1721	Sodium acetate
184.1724	Sodium alginate
184.1733	Sodium benzoate
184.1736	Sodium bicarbonate
184.1742	Sodium carbonate
184.1754	Sodium diacetate
184.1763	Sodium hydroxide
184.1764	Sodium hypophosphite
184.1769a	Sodium metasilicate
184.1792	Sodium sesquicarbonate

184.1784   Sodium propionate
184.1801   Sodium tartrate
184.1804   Sodium potassium tartrate
184.1807   Sodium thiosulfate
184.1835   Sorbitol
184.1845   Stannous chloride (anhydrous and dihydrated)
184.1848   Starter distillate
184.1875   Thiamine hydrochloride
184.1878   Thiamine mononitrate
184.1890 α-Tocopherols
184.1923   Urea
184.1930   Vitamin A
184.1945   Vitamin B_{12}
184.1950   Vitamin D
184.1973   Beeswax (yellow and white)
184.1976   Candelilla wax
184.1978   Carnauba wax
184.1979   Whey
184.1979a Reduced lactose whey
184.1979b Reduced minerals whey
184.1979c Whey protein concentrate
184.1983   Bakers yeast extract
184.1984   Zein

Part 186. Indirect Food Substances Affirmed
As Generally Recognized As Safe

These substances are added indirectly to human food for the purpose and under the conditions prescribed, providing the substances comply with the listed purity specifications or are of a purity suitable for their intended use. These substances have been reviewed and are generally recognized as safe.

Subpart A — General Provisions
186.1   Substances added indirectly to human food affirmed as generally recognized as safe (GRAS)

Subpart B — Listing of Specific Substances Affirmed as GRAS
186.1025   Caprylic acid
186.1093   Sulfamic acid
186.1256   Clay (kaolin)
186.1275   Dextrans

186.1316   Formic acid
186.1551   Hydrogenated fish oil
186.1557   Tall oil
186.1673   Pulp
186.1750   Sodium chlorite
186.1756   Sodium formate
186.1770   Sodium oleate
186.1771   Sodium palmitate
186.1797   Sodium sulfate
186.1839   Sorbose

Bibliography

Adipic Acid. St. Louis, Missouri: Monsanto.

Adipic Acid. Product Bulletin. Dallas, Texas: Celanese Chemical Co., Inc.

Alberger Salt for Food Processing. St. Clair, Michigan: Diamond Crystal Salt Co.

Aluminum Sulfate. Technical Bulletin. Morristown, New Jersey: Allied Chemical Company.

Altschul, A. *Protein Forum.* Decatur, Illinois: A. E. Staley Mfg. Co.

Amaizo Corn Syrups. Hammond, Indiana: American Maize-Products Co.

Amaizo Food Specialty Starches. 1976. Hammond, Indiana: American Maize-Products Co.

Amaizo Fro-Dex® Corn Syrup Solids. 1979. Hammond, Indiana: American Maize-Products Co.

American Dairy Products Institute. 1987. Select the Whey Product to Meet Your Ingredient Needs. *Food Technology* 41(10): 124–125.

Andres, C. 1988. Antioxidants—'Quality Protectors.' *Food Processing* 46(2): 36–41.

Andres, C. 1984. Black Pepper—The Versatile Flavor. *Food Processing* 45(5): 84–86.

Andres, C. 1980. Corn—A Most Versatile Grain. *Food Processing* 41(5): 78–100.

Andres, C. 1982. Food Type, Operating Procedures, Food pH Enter into Color Decision. *Food Processing* 43(5): 52–53.

Andres, C. 1984. Gluten—The Versatile Ingredient. *Food Processing* 45(5): 80–82.

Andres, C. 1982. Mold/Yeast Inhibitor Gains FDA Approval for Cheese. *Food Processing* 43(11): 83.

215

Andres, C. 1975. Processors Guide to Gums — Part I. *Food Processing* 36(12): 35 – 36.

Andres, C. 1976. Stabilizers 2 — Gums. *Food Processing* 37(1): 83 – 87.

Andres, C. 1984. Sulfites. *Food Processing* 46(3): 34.

Andres, C. 1976. Surfactants/Emulsifiers Improve Texture, Quality, Shelf Life of Foods. *Food Processing* 37(5): 63 – 68.

Andres, C. 1984. Three Factors Affecting Price/Supply. *Food Processing* 45(4): 77.

Andres, C. 1982. Yellow and Reddish-Orange Colors Have Good Stability and Solubility. *Food Processing* 43(5): 44 – 45.

Annatto Colors. Technical Bulletin. Milwaukee, Wisconsin: Chr. Hansen's Laboratory, Inc.

Annatto Colors. Technical Bulletin. Elkhart, Indiana: Miles Laboratories, Inc.

Anti-caking and Conditioning Agents. 1977. Edison, New Jersey: J. M. Huber Corp.

Apple, The. The Pectin. Technical Bulletin. Neuenbürg, FR Germany: Herman Herbstreith KG.

Aromatics. Technical Bulletin. Los Angeles, California: F. Ritter & Co.

Ascorbyl Palmitate NF,FCC. Product Data Bulletin. Nutley, New Jersey: Roche Chemical Division.

Atlas Food Color Guide. New York: H. Kohnstamm & Co., Inc.

Altas Products for Foods. Wilmington, Delaware: ICI United States, Inc.

Avicel® PH 101 in Grated and Shredded Cheese. Technical Bulletin. Philadelphia: FMC Corp.

Avicel® PH Microcrystalline Cellulose. Technical Bulletin. Philadelphia: FMC Corp.

Avicel® RC. Technical Bulletin. Philadelphia: FMC Corp.

Bacteria, Yeast & Molds. 1972. Athens, Georgia: Cooperative Extension Service, University of Georgia.

Barley. Cannon Falls, Minnesota: Minnesota Grain Pearling Company.

Benzoic Acid & Sodium Benzoate. 1970. St. Louis, Missouri: Monsanto.

Calcium Lactate. Gorinchem, Holland: C. V. Chemie Combinatie Amsterdam.

Calcium Sulfate for Food and Pharmaceutical Uses. Technical Bulletin. Chicago: United States Gypsum.

Calcium Sulfate for the Baking Industry. 1978. Technical Bulletin. Chicago: United States Gypsum.

Cal Plus™. Technical Bulletin. St. Louis, Missouri: Mallinckrodt, Inc.

Cal Tac™. Technical Bulletin, St. Louis, Missouri: Mallinckrodt, Inc.

Caramel Coloring. Cedar Rapids, Iowa: Corn Sweeteners.

Cassidy, J. P. 1977. Phosphates in the Dairy Industry. *Food Product Development* 11(5): 83 – 85.

Cerelose® Dextrose. 1980. Englewood Cliffs, New Jersey: CPC International Inc.

Certified FD&C Colors. 1979. Technical Bulletin. Cincinnati, Ohio: Hilton-Davies.

Chemicals Used in Food Processing. 1965. Washington, D.C.: National Academy of Sciences-National Research Council.

Chedd, G. 1974. The Search for Sweetness. *New Scientist.* (May): 299–302.

Chocolate Information. Technical Bulletin. Hunt Valley, Maryland: McCormick & Co., Inc.

Citric Acid. Technical Bulletin No. 78. New York: Pfizer Chemicals Division.

Code of Federal Regulations, Title 21, Parts 1 to 99, Parts 100–169, Parts 170–199. Washington, D.C.: Superintendent of Documents. Government Printing Office (published annually).

Comparison of the Properties of Pure Crystalline Fructose and Isomerized Corn Syrups, A. Technical Bulletin. Nutley, New Jersey: Roche Chemical Division.

Corn Bran-Regular. Technical Bulletin. Paris, Illinois: Illinois Cereal Mills, Inc.

Creative Cooking with Bulgur Wheat. Danville, Illinois: Lauhoff Grain Company.

Current FDA Status of Atlas® Surfactants and Polyols for Use in Foods. Technical Bulletin. Wilmington, Delaware: ICI United States Inc.

CWS Fumaric Acid. St. Louis, Missouri: Monsanto.

Damon, G. E. 1974. A Primer on Vitamins. *FDA Consumer* (May): 5–11.

Dehydroacetic Acid. Publication No. A-106A. 1980. Kingsport, Tennessee: Eastman Chemical Products, Inc.

Delvocid®. Product Bulletin Del-03/82.10 Am.10. Charlotte, North Carolina: G. B. Fermentation Industries Inc.

D-Fructose, Food Grade. Technical Bulletin. Nutley, New Jersey: Roche Chemical Division.

Dietary Salt. 1980. *Food Technology* 34(1): 85–90.

Distilled Monoglyceride Type S, Type S (V). Tokyo, Japan: Riken Vitamin Oil Co., Ltd.

Durkee Product Catalog. 1979. Cleveland, Ohio: Durkee Industrial Foods Group.

Durkee Shortening & Oil Glossary. Cleveland, Ohio: Durkee Industrial Foods Group.

Durkote GDL 135-50. 1978. Technical Bulletin. Cleveland, Ohio: Durkee Industrial Foods Group.

Earl, E, and J. F. Rand. 1977. Foam Control in Potato Processing. *Food Engineering* 49(11): 71–73.

Eastman Triacetin, Food Grade. Technical Bulletin. Kingsport, Tennessee: Eastman Chemical Products, Inc.

Edifas® A Methyl Ethyl Cellulose. 1974. Technical Bulletin. Wilmington, Delaware: ICI America, Inc.

Edible Rennet Casein. New Zealand: New Zealand Dairy Board.

Eggs in Brief. 1985. Park Ridge, Illinois: The American Egg Board.

Eggs, Your Diet and Your Health. Park Ridge, Illinois: National Commission on Egg Nutrition.

Emodi, A. 1978. Xylitol, Its Properties and Food Applications. *Food Technology* 32(1): 28–32.

Engineered Ingredients: Building Blocks for Food Design. 1975. *Food Engineering* 47(7): 46–47.

Enocyanin. Technical Bulletin. Plainview, New York: Burlington Bio-Medical & Scientific Corporation.

Erythorbic Acid and Sodium Erythorbate in Foods. 1975. New York, New York: Pfizer Chemicals Division.

Esselen, W. B. 1975. Applications for Rosells as a Red Food Colorant. *Food Product Development.* 9(8): 37–40.

Essential Oils and Oleoresins. Madison, Wisconsin: Quality Control Spice Co., Inc.

"Fagopyrum sagittatum." Or "buckwheat." 1980. Cannon Falls, Minnesota: Minnesota Grain Pearling Company.

FDA Approves Acesulfame-K for Use in Foods and Beverages. 1988. *Food Engineering International* 13(8): 16.

Flavor Enhancers and Nutritional Ingredients. Technical Bulletin. Clifton, New Jersey: Yeast Products Inc.

Flour. Manhattan, Kansas: American Institute of Baking.

FM Custom Flavors. Chicago, Illinois: Food Materials Corporation.

Foamkill Silicone Antifoam Compounds. Technical Bulletin. Greenville, South Carolina: Crucible Chemical Co.

Food Acidulants. New York: Pfizer Chemicals Division.

Food Chemical New Guide, The. Washington, D.C.: Louis Rothchild, Jr.

Food Chemicals Codex. 1972. 2nd ed. Washington, D.C.: National Academy of Sciences.

Food Emulsifiers. Greenwich, Connecticut: Glyco Chemicals Inc.

Food Emulsifiers. Gurnee, Illinois: Mazer Chemicals, Inc.

Food Emulsifiers, Coatings, and Lubricants. 1976. Kingsport, Tennessee: Eastman Chemical Products, Inc.

Food Emulsifiers and Specialty Products. 1977. Boonton, New Jersey: PVO International Inc.

Food Fortification Aids. 1982. *Food Processing* 43(10): 48.

Food Grade Lactic Acid. St. Louis, Missouri: Monsanto.

Food Grade Phosphates, Phosphoric Acid. Philadelphia: FMC Corporation.

Food Processing Applications of "Purity Brand" Sodium Caseinate. Technical Bulletin. Erie, Illinois: Erie Casein Co., Inc.

Food Preservatives. New York: Pfizer Chemicals Division.

Food Service Seasoning Guide. New York: American Spice Trade Association.

Fructose Fruit Sugar. Helsinki, Finland: Oy Finnsugar Trading Co.

Fumaric Acid. 1977. St. Louis, Missouri: Monsanto.

Functions of Phosphates in Foods. Philadelphia: FMC Corporation.

Furia, T. E. 1968. *Handbook of Food Additives.* Cleveland, Ohio: The Chemical Rubber Co.

Gluconate. Technical Bulletin. The Netherlands: Akso Chemie Nederland B.V.

Gluconic Acid. West Germany: Joh. A. Benckiser GmbH.

Goldschmiedt, H. 1974. Why Are Anti-Oxidants Needed in the Food Industry? *Food Trade Review* (March): 10–12.

GPC Bulletin No. 1004. Muscatine, Iowa: Grain Processing Corporation.

Graf, T., and L. Meyer. 1976. Use of Lecithin in Food. *International Flavors and Food Additives* 7: 218–21.

Grindsted Emulsifiers. Overland Park, Kansas: Grindsted Products, Inc.

Griswold, R. M. 1962. *The Experimental Study of Foods.* Boston: Houghton Mifflin Co.

Glycine. Technical Bulletin. Lexington, Massachusetts: W. R. Grace & Company.

Hannigan, K. J. 1980. Crystalline Fructose: Who's Using It and Why. *Food Engineering* 52(11): 28–30.

Hannigan, K. J. 1980. Spices: Changes Ahead. *Food Engineering* 52(6): 47.

Hardy High Grade Salt for the Food Industry. St. Louis, Missouri: Hardy Salt Co.

Hawaiian Washed Raw Sugar. Bulletin. San Francisco, California: C&H Sugar Company.

Hawley, G. G. 1977. *The Condensed Chemical Dictionary.* 9th ed. New York: Van Nostrand Reinhold Co.

Henderson, J. L. 1971. *The Fluid-milk Industry.* Westport, Connecticut: AVI Publishing Co., Inc.

Houston, D. F. *Rice.* St. Paul, Minnesota: American Association of Cereal Chemists.

Hydrogenated Glucose Syrup Permits New Sugar-Free Candies. 1984. *Prepared Foods* 153(3): 158.

Hysorb®-40. Technical Data. Denver, Colorado: Manville Corporation.

ICP Cocoa Inc. Camden, New Jersey: ICP Cocoa Inc.

Idea Bank Book, The. Kansas City, Missouri: Patco Products, C.J. Patterson Co.

Index to Mere-Colors. Technical Bulletin. North Bergen, New Jersey: Meer Corporation.

Information About Silicon Antifoam. Medland, Michigan: Dow Corning Corp.

Ingredients for Food Processing. St. Louis, Missouri: Monsanto.

Ingredients Handbook. 1979. *Food Processing* 40(5): 133–151.

Iron. 1976. Data Sheet. St. Louis, Missouri: Mallinckrodt Food Products.

Iron Sources. Technical Bulletin. St. Louis, Missouri: Mallinckrodt Food Products.

Is There Corn in Your Product's Future? Englewood Cliffs, New Jersey: Corn Products.

Isosweet® 100 Fructose Corn Syrup. Technical Bulletin. Decatur, Illinois: A. E. Staley Mfg. Co.

Jones, N. R. 1972. Natural Stabilizers. *Food Technology* 24(12): 626–639.

Kroger, M. 1972, October. Controlling the Quality of Yogurt. Paper presented at Dairy Quality Control Conference.

Lachance, P. A. 1973. Carbohydrates as Nutrients. *Food Product Development* 7(6): 29–34.

Lactate Esters. Technical Bulletin. Holland: C. V. Chemie Combinate Amsterdam C.C.A.

Lactic Acid and Lactates. Gorinchem, Holland: C. V. Chemie Combinatie. Amsterdam C.C.A.

Lactic Acid and Lactates. Technical Bulletin. London, England: Croda Food Products Ltd.

Lactose. 1979. San Francisco, California: Foremost-McKesson Inc.

Lactose in Bakery Products. 1977. Technical Bulletin. San Francisco, California: Foremost-McKesson, Inc.

Lampert, L. M. 1975. *Modern Dairy Products.* New York: Chemical Publishing Co., Inc.

Leciflow. Technical Bulletin. Chicago: Kraft Foods.

Lecithin. Technical Bulletin. Chicago: Central Soya Co., Inc.

Lecithin from Central Soya Naturally. Fort Wayne, Indiana: Central Soya Co., Inc.

Lehmann, P. 1979, April. More Than You Ever Thought You Would Know About Food Additives. *FDA Consumer.*

Lonzest Polyethoxylated Sorbitan Esters. Technical Bulletin. Fair Lawn, New Jersey: Lonza Inc.

Low Calorie Sweeteners. 1983. *Food Engineering* 55(5): 138–139.

McCormick, R. D. 1975. Aspartame: A New Dimension for Controlling Product Sweetness. *Food Product Development* 9(1): 22, 35–36.

McCormick, R. D. 1983. Food Pigment Use Governed by Broad Considerations. *Processed Prepared Foods* 152(1): 127, 131–132.

McCormick, R. D. 1986. Ginger—The Zing in Asian Cuisine. *Processed Prepared Foods* 155(6): 169–171.

McCormick, R. D. 1982. GRAS Emulsifier Has Broad Spectrum Antimicrobial Properties. *Processed Prepared Foods* 151(8): 125.

McCormick, R. D. 1985. Methods for Developing Unique Textured Foods. *Prepared Foods* 154(3): 155–156.

McDonald, R. W. *Salt.* Chicago: Morton Salt.

Make Good Foods Better. Technical Bulletin. Fort Wayne, Indiana: Central Soya Co., Inc.

Malic Acid. Parsippany, New Jersey: Alberta Gas Chemicals.

Maltrin®. Muscatine, Iowa: Grain Processing Corp.

Marks, J. 1968. *The Vitamins in Health and Disease.* London, England: J&A Churchill Ltd.

Matz, S. A. *Bakery Technology And Engineering.* Westport, Connecticut: The AVI Publishing Co., Inc.

Matz, S. A. *Cookie and Cracker Technology.* Westport, Connecticut: The AVI Publishing Co., Inc.

Matz, S. A. 1959. *The Chemistry and Technology of Cereal as Food and Feed.* Westport, Connecticut: The AVI Publishing Co., Inc.

Maxarome®. Des Plaines, Illinois: GB Fermentation Industries, Inc.

Meer Natural Colors. Technical Bulletin. North Bergen, New Jersey: Meer Corp.

Minerals for Nutrition. Technical Bulletin. St. Louis, Missouri: Mallinckrodt, Inc.

Monsanto Sorbic Acid and Potassium Sorbate. St. Louis, Missouri: Monsanto.

Morris, C. E. 1981. Combines Intense Flavors With Low Salt Content. *Food Engineering* 53(9): 94–95.

Morris, C. E. 1981. FDA Clears Aspartame. *Food Engineering* 53(8): 154–155.

National Certified FD&C Food Colors. 1972. Morristown, New Jersey: Allied Chemical Corporation.

Natural Fats and Oils. Technical Bulletin. Boonton, New Jersey. Drew Chemical Corp.

Natural Spices. Madison, Wisconsin: Quality Control Spice Co., Inc.

Natural Vegetable Colors. Technical Bulletin. Milwaukee, Wisconsin: Chr. Hansen's Laboratory, Inc.

Nelson, J. A., and G. M. Trout. 1964. *Judging Dairy Products.* Milwaukee, Wisconsin: Olsen Publishing Co.

Newer Knowledge of Milk. 3rd ed. Chicago: National Dairy Council.

New Preservative Proven Effective in Tests. 1982. *Food Engineering* 54(11): 60.

New Trade-Offs in Selecting Acidulants. 1974. *Food Engineering* 46(5): 84–85.

OBI Pectin. Technical Bulletin. Switzerland: OBI Pectin AG.

Ockerman, H. W. 1978. *Source Book for Food Scientists.* Westport, Connecticut: The AVI Publishing Co., Inc.

Official Methods of Analysis. 1975. 12th ed. Washington, D.C.: Association of Official Analytical Chemists.

Oleoresin Handbook, The. 1974. New York: Fritzsche Dodge & Olcutt Inc.

On Lactose. San Francisco, California: Foremost-McKesson, Inc.

Paul, P. C., and H. H. Palmer. *Food Theory and Applications.* New York: John Wiley & Sons, Inc.

Peanut Oil. Tifton, Georgia: Georgia Agricultural Commodity Commission for Peanuts.

Peckham, G. C. 1964. *Foundations of Food Preparation: A Beginning College Text.* New York: The Macmillan Company.

Penick Plus-A-Tives® and Flavor Materials. New York: S. B. Penick & Co.

Petrowski, G. E. 1976. Emulsion Stability and Its Relation to Foods. Reprinted from *Advances in Food Research* 22.

Petrowski, G. E. 1975. Food-Grade Emulsifiers. *Food Technology* 29(7): 52–62.

Phosphates. 1979. Technical Bulletin. Philadelphia, Pennsylvania: FMC Corporation.

Phosphoric Acid. Westport, Connecticut: Stauffer Chemical Co.

Polydextrose. 1982. New York: Pfizer, Inc.

Pomalus. 1977. Morristown, New Jersey: Allied Chemical Corp.

Potassium Chloride. Westwood, New Jersey: Tri-K Industries, Inc.

Potassium Hydrogen Tartrate. FR Germany: Joh. A. Benckiser GmbH.

Product Catalog 1977–1978. Harbor City, California: Hathaway Allied Products, Inc.

Product Guide Cocoa Powder. 1974. Holland: Cacaofabriek de Zaan.

Product Information on Phosphates. Technical Bulletin. Westport, Connecticut: Stauffer Chemical Company.

Promine®-F. Chicago: Central Soya Co., Inc.

Promine®-R. Chicago: Central Soya Co., Inc.

Promosoy®-100. Chicago: Central Soya Co., Inc.

Propylene Glycol. New York: Union Carbide Corp.

Protein Power of Soybeans, The. Chicago: Food Protein Council.

Proteins from EEC. Erie, Illinois: The Erie Casein Company, Inc.

Przybyla, A. Colors. 1980. *Processed Prepared Food* 149(9): 109–114.

Przybyla, A. 1983. Oils: Oil Selection Key to Many New Products. *Prepared Foods* 152(4): 97–100.

Przybyla, A. 1981. Polydextrose New Food Additive Approved By FDA. *Processed Prepared Foods* 150(8): 75–76.

Pyridoxine Hydrochloride. Technical Bulletin. New York: Fallek Sales Inc.

Questions and Answers About Phosphates in Food. Westport, Connecticut: Stauffer Chemical Company.

Remarkable Story of Monosodium Glutamate, The. Washington, D.C.: International Glutamate Technical Committee.

Riboh, M., 1977. Natural Colors: What Works . . . What Doesn't. *Food Engineering* 49(5): 67–72.

Ribotide®. Tokyo: Takeda Chemical Industries, Ltd.

Rice, J. 1983. Yucca Plant Extract. *Food Processing* 44(4): 58.

Robinson, R. F. 1972. What is the Future of Textured Protein Products? *Food Technology* 26(5): 59–63.

Roche Carotenoids. 1986. Technical Bulletin. Nutley, New Jersey: Roche Chemical Division.

Rust, R. E., and D. G. Olson. *Meat Curing Principles and Modern Practice.* Kansas City, Missouri: Koch Supplies, Inc.

SAG® Silicone Antifoams. 1975. New York: Union Carbide.

Sales Soar for Carob Powder Used in Candies, Cookies, Cakes, Instant Drinks . . . 1977. *Food Engineering* 49(9): 43.

Salt in the Meat Industry. St. Clair, Michigan: Diamond Crystal Salt Co.

Saussele, H. Jr. 1982. Trends and Future of Yeast. *Food Engineering* 54(5): 129.

Sequestrene Na₂ Food Grade and Na₂Ca Food Grade EDTA. 1985. Greensboro, North Carolina: Ciba-Geigy Corporation.

Signorino, C. A., and E. R. Furmanski. 1975. Dyes Cause Color Problems? Try Certified Lakes. *Food Engineering* 47(5): 76–77.

Silicone Antifoams. 1973. Skokie, Illinois: Hodag Chemical Corp.

Sinclair, P., R. S. Vettel, and C. A. Davis. *Soybeans in Family Meals.* Washington, D.C.: U.S. Dept. of Agriculture.

Sodium Aluminum Phosphate, Basic. Technical Bulletin. Westport, Connecticut: Stauffer Chemical Co.

Sodium Bicarbonate. 1987. New York: Church & Dwight Co., Inc.

Sodium Bicarbonate. 1982. Technical Bulletin. Morristown, New Jersey: Allied Chemical.

Sodium Citrate in Foods. 1969. New York: Pfizer Chemicals Division.

Sodium Saccharin Powder. 1981. Chicago: Pettibone-Chicago, Inc.

Solka-Floc®. New York: Brown Company.

Sorbic Acid and Potassium Sorbate. St. Louis, Missouri: Monsanto.

Sorbitol. 1972. Technical Bulletin. Fairlawn, New Jersey: Lonza, Inc.

Sorbitol in Foods. 1975. New York: Pfizer Chemicals Division.

Sorbitol, Sorbitol. 1971. New York: Pfizer Chemicals Division.

Specification for the Identity and Purity of Food Additives and Their Toxicological Evaluation: Emulsifiers, Stabilizers, Bleaching and Maturing Agents. 1964. Geneva, Switzerland: World Health Organization.

Spices, Flavors, Flavor Characterization. 1966. New York: Fritzche Dodge & Olcott Inc.

Starch. Technical Bulletin. Bridgewater, New Jersey: National Starch and Chemical Company.

Starch. Technical Data. Decatur, Illinois: A. E. Staley Mfg. Co.

Star-Dri® Corn Syrup Solids. Decatur, Illinois: A. E. Staley Mfg. Co.

Sta-Sol® R Lecithin Compound. Technical Bulletin. Decatur, Illinois: A. E. Staley Mfg. Co.

Sta-Sol® Lecithin Concentrates. Technical Bulletin. Decatur, Illinois: A. E. Staley Mfg. Co.

Stephenson, M. G. 1975. Textured Plant Protein Products: New Choices for Consumers. *FDA Consumer* 9(3): 18–23.

Straight Talk About Salt. 1982. Alexandria, Virginia: Salt Institute.

Strike Oil. Bismarck, North Dakota: North Dakota Sunflower Council.

Sugarless Syrup. 1984. *Food Engineering* 56(5): 68–69.

Sunflower. 1982. Bismarck, North Dakota: National Sunflower Association.

Sustane® Food-Grade Antioxidants. Technical Bulletin. Des Plaines, Illinois: UOP Process Division.

Syloid®. Baltimore, Maryland: W. R. Grace & Co.

Syloid® Conditioning Agents for the Food Industry. Baltimore, Maryland: W. R. Grace & Co.

Syncal™ Calcium Saccharin. 1981. Cleveland, Ohio: Sherwin-Williams Company.

Syncal™ Sodium Saccharin. 1981. Cleveland, Ohio: Sherwin-Williams Company.

Takeda-Fallek Sales, Inc. Technical Bulletin. New York: Takeda-Fallek Sales Inc.

Tannic Acid. Technical Bulletin. Flushing, New York: Aceto Chemical Co., Inc.

Tenox® Food-Grade Antioxidants. 1974. Kingsport, Tennessee: Eastman Chemical Products.

Terra Alba, Calcium Sulfate Dihydrate. Technical Bulletin. Elkhart, Indiana: Miles Laboratories, Inc.

Testing Time for Tofu. 1984. *Food Manufacture* 59(5): 76–77.

Triebold, H. O., and L. W. Aurand. 1969. *Food Composition and Analysis.* New York: Van Nostrand Reinhold Co.

Tripotassium Citrate. FR Germany: Joh. A. Benckiser GmbH.

Trisodium Citrate. FR Germany: Joh. A. Benckiser GmbH.

Twigg, B. A. 1974. *Ingredient Technology for Product Development.* Portland, Oregon: IFT Short Course.

Your Guide for Buying Malt Extracts and Related Syrups. Milwaukee, Wisconsin: Premier Malt Products, Inc.

Valley Evaporating Company. Specification Sheet #21. Yakima, Washington: Kingblossom Apple Flakes & Powders.

Veltol®-Plus and Veltol®. New York: Pfizer Chemicals Division.

Versene® Ca and Versene® NA Food Grade Chelating Agents. Technical Bulletin. Midland, Michigan: Dow Chemical Co.

Wheat Gluten. 1981. Shawnee Mission, Kansas: International Wheat Gluten Association.

Wheat Starch. Technical Bulletin. Minneapolis, Minnesota: Henkel Corp.

Whistler, R. L., and J. N. Bemiller. 1973. *Industrial Gums.* New York: Academic Press.

Windholz, M. 1976. *The Merck Index.* 9th ed. Rahway, New Jersey: Merck & Co., Inc.

Woodin, G. B. 1981. The Elegant Egg. *Food Product Development* 15(4): 44–56.

Worcestershire Sauce. Technical Bulletin. Dallas, Texas: The Illes Co.

World of Norda Flavors, The. New York: Norda Essential Oil & Chemical Co., Inc.

Xylitol: Sugar's New Rival. 1976. *Business Week*, September 3, p. 765.

Yeasts for Wine Making. 1981. Bulletin. Milwaukee, Wisconsin: Universal Foods Corporation.

Ziemba, J. V. 1975. Food Protein Research Accelerates. *Food Processing* 36(8): 21–27.

Zoller, J. M. et al. 1980. Fortification of Non-Staple Food Items with Iron. *Food Technology* 34(1): 38–46.